Concept Building in Fisheries Data Analysis

Basant Kumar Das • Dharm Nath Jha •
Sanjeev Kumar Sahu • Anil Kumar Yadav •
Rohan Kumar Raman • M. Kartikeyan

Concept Building in Fisheries Data Analysis

Springer

Basant Kumar Das
ICAR
Central Inland Fisheries Research Institute
Barrackpore, West Bengal, India

Dharm Nath Jha
Prayagraj Regional Centre
ICAR-CIFRI
Prayagraj, Uttar Pradesh, India

Sanjeev Kumar Sahu
Fisheries Resource Assessment & Informatics
ICAR-CIFRI
Barrackpore, West Bengal, India

Anil Kumar Yadav
Guwahati Regional Centre
ICAR-CIFRI
Gowahati, Assam, India

Rohan Kumar Raman
Division of Socio Economics & Extension
ICAR Research Complex for Eastern Region
Patna, Bihar, India

M. Kartikeyan
Bengaluru Regional Centre
ICAR-CIFRI
Bangalore, Karnataka, India

ISBN 978-981-19-4410-9 ISBN 978-981-19-4411-6 (eBook)
https://doi.org/10.1007/978-981-19-4411-6

This work is a co-publication with Narendra Publishing House, New Delhi. This work is licensed for electronic sale in all countries. The work is licensed for print sale in all countries except India. Sold and distributed in print in India by Narendra Publishing House C-21, Varun Apartment, Sector-9, Rohini, DELHI – 110085, India
Jointly published with Narendra Publishing House
ISBN of the Narendra Publishing House edition: 9789390611676

This Springer imprint is published by the registered company Springer Nature Singapore Pte Ltd.
The registered company address is: 152 Beach Road, #21-01/04 Gateway East, Singapore 189721, Singapore

Foreword

सत्यमेव जयते
त्रिलोचन महापात्र, पीएच.डी.
सचिव, एवं महानिदेशक
TRILOCHAN MOHAPATRA, Ph.D.
SECRETARY & DIRECTOR GENERAL

भारत सरकार
कृषि अनुसंधान और शिक्षा विभाग एवं
भारतीय कृषि अनुसंधान परिषद
कृषि एवं किसान कल्याण मंत्रालय, कृषि भवन, नई दिल्ली 110 001

GOVERNMENT OF INDIA
DEPARTMENT OF AGRICULTURAL RESEARCH & EDUCATION
AND
INDIAN COUNCIL OF AGRICULTURAL RESEARCH
MINISTRY OF AGRICULTURE AND FARMERS WELFARE
KRISHI BHAVAN, NEW DELHI 110 001
Tel.: 23382629; 23386711 Fax: 91-11-23384773
E-mail: dg.icar@nic.in

India has emerged as one of the largest producers of fish in the world with a total production of 13.76 mmt, contributing 7.58% to global production. Of the total fish production, over 70% comes from the Inland sector. The Inland fisheries resources are vast in the form of rivers, canals, floodplain lakes, ponds, reservoirs and brackish water bodies. These resources are bestowed with wide varieties of fish species, which are distributed based on their physiological and climatic requirements. The fisheries sector and its spin-off activities provide livelihood opportunities to a diverse group of rural artisans. Mapping of the resources, fish distribution pattern, associated fishing communities, methods of fishing, income generating opportunities, trade, etc. has led to the generation of a colossal volume of data. To analyse and interpret this data, there is a need to have various data analysis tools.

I am very happy that the ICAR-Central Inland Fisheries Research Institute, Barrackpore, is publishing the book entitled *Concept Building in Fisheries Data Analysis*. The book builds up a conceptual foundation of frequently used statistical data analysis tools, covering both theoretical and practical approaches and its analysis with various standard software packages. The most important part of the book is data interpretation from the analysis output, which, in my opinion, will be very helpful to the practitioners involved in fisheries. I am pleased with the structural design of the book, especially the way it has demonstrated the statistical methods by using real-life fisheries data. I sincerely believe that this book will help students, practitioners and teachers associated with inland fisheries and guide them to perform fundamental statistical analyses along with meaningful interpretation of results.

I compliment the authors for their stupendous efforts in bringing out this book—a veritable treatise on both theoretical and practical concepts of statistics in one place. I wish them all the very best for their future endeavours.

Indian Council of Agricultural Research
New Delhi, India

T. Mohapatra

Foreword

Fisheries professionals—most notably, students and researchers—invariably provoke a response to statistical data analysis as 'Statistics was the worst subject I liked in my student life'. This negative attitude towards statistics consequently makes them fumble in real-life fisheries data analysis. I do know the reason that most of them cannot afford to fulfil: the need to be competent in statistics. I believe a good book can help to come out from this dogma, and competency can be acquired, as far as basic data analysis is concerned. Unfortunately, statistics books exclusively dealing with fisheries data analysis are rare—especially in India.

I am overwhelmed by reading this book entitled *Concept Building in Fisheries Data Analysis* brought out by the group of statisticians at ICAR-Central Inland Fisheries Research Institute, Barrackpore. The book is well structured, starting from field data collection methods, analysis and to the interpretation of results. Though the mathematical formulas are imperative in statistics, this book has emphasized bringing clarity in concept, which will build a strong base in statistics for students. Books on data analysis are undeniably plentiful, but they describe statistics in general. But, this book is unique because of its exclusive fisheries data analysis and subsequent interpretation thereof. The data presentation described in the book is impressive. The authors make the readers feel that statistics is away from mathematical complexities and could be easy to understand, appealing and interesting. The strongest part of the book is presentation of examples, which are taken exclusively from field-based fisheries interventions with their post-analysis interpretations. The exercise provided in each chapter will be highly useful for students. Demonstration software—Excel and R—introduced in the book is the prudent choice. Fisheries professionals feel at home in excel while R is open-source, incurring practically no cost.

The authors indeed deserve appreciation for bringing out this book. I wish them all the success of their endeavour and will be looking forward to reading more such books by the authors.

Assam Fisheries Development Corporation Dilip Kumar
Guwahati, India

National Platform for Small Scale
Fish Workers (NPSSFW)
Kolkata, India

Board of Governors,
Institute of Livelihood Research
and Training (ILRT)
Hyderabad, India

ICAR-CIFE
Mumbai, India

International Civil Service (FAO of UN)
New York, NY, USA

Preface

Effective management strategies of inland fisheries demand efficient quantitative data analysis and subsequent interpretation of results. The contribution of statistics is imperative wherever data is subject to uncertainty which is frequently encountered in generating inland fisheries data. Thereby, estimation, interpretation and prediction of fishery resources that lead to formulating government policy or any other related issue, and the role of statistics comes in.

Thus it is necessary that an updated and improved understanding of the statistical concepts is essential for students, researchers, planners, decision-makers and officials associated with inland fisheries. With this in mind, a first attempt was made in 2018, to provide training supported by a training manual to help the researchers associated with inland fisheries. With the success and suggestions received from the participants of first training, an improvised manual was prepared during the second training programme held in 2019. Based on the feedback of the trainees, the authors conceived the idea of publishing a book. The book design emphasizes on building the foundation on the statistical concept so that readers can easily understand and execute fundamental statistical analyses themselves. It has both theoretical concept and practical implementation of that theory. The theoretical part aims at building and explaining concepts whereas the practical sessions focus primarily on a disciplined way of data entry followed by analyses by software. Special emphasis has been given on data interpretation.

It is very challenging to build up key concepts of statistics without a mathematical background. Detailed mathematical derivation of the methods—especially for multivariate analysis—has been avoided purposefully because of the targeted audience who are mainly the users of statistics. Therefore, the authors have provided an essential reference list for advanced users. Utmost care has been taken to demonstrate the statistical methods by using real-life fisheries data. The authors hope that this book will be helpful to perform frequently used statistical analyses and to interpret the results in a meaningful way. Although sophisticated analyses are appealing, the necessary foundation in statistics is essential to proceed towards high-level analysis. The authors hope that this book is an essential step in that direction.

Barrackpore, West Bengal, India	Basant Kumar Das
Prayagraj, Uttar Pradesh, India	Dharm Nath Jha
Barrackpore, West Bengal, India	Sanjeev Kumar Sahu
Gowahati, Assam, India	Anil Kumar Yadav
Patna, Bihar, India	Rohan Kumar Raman
Bangalore, Karnataka, India	M. Kartikeyan

Acknowledgements

Foremost, the authors are enormously grateful to the scientists, and trainees who underwent training courses organized at the ICAR-Central Inland Fisheries Research Institute—without their feedback and request the book would not have been possible. The authors express their gratitude to the Director, ICAR-Central Inland Fisheries Research Institute, Barrackpore, for rendering the support and encouragement.

The authors also wish to express their profound sense of thankfulness and indebtedness to Dr. Malay Naskar, Principal Scientist, ICAR-CIFRI, Barrackpore, for his earnest endeavour, intellectual guidance and valuable suggestions in bringing out this book. Without his support and toil, it would not have been possible to publish this book in the present format.

Statistical analysis does not exist without data. There are a lot of illustrative real data examples, without which the book would not be helpful for students and practitioners. The authors owe enormously to Dr. S. K. Manna, Dr. Raju Baitha, Dr. Sona Yengkokpam and Assam Fisheries Development Corporation (AFDC) for sharing their data used for demo analysis in this book. The authors also sincerely acknowledge the significant contribution of Mohammed Naim who assisted with typesetting using LaTeX.

Authors

Contents

1 Nature and Sources of Data for Inland Fisheries Management 1
1.1 Complexity of Inland Fisheries Management 2
1.2 Data Requirement 3
1.2.1 Objectives 3
1.2.2 Data Type 4
1.2.3 Data and Information 4
1.2.4 Location and Time 7
1.3 Data Collection Methods 8
1.4 Some Frequently Collected Data in Inland Fisheries 9
1.4.1 Data for Biotic–Environment Assessment 10
1.4.2 Spatial Data for Fishery Assessment 12
1.4.3 Data for Stock Enhancement Strategies 12
1.4.4 Survey Data 13
1.4.5 Perception-Based Data 13
1.5 Conclusion 14
Further Reading 15

2 Concept of Sampling Methodologies and Their Applications 17
2.1 Some Definitions 18
2.2 Simple Random Sampling 19
2.3 Systematic Sampling 22
2.3.1 Advantages and Disadvantages Advantage 22
2.4 Stratified Random Sampling 24
2.4.1 Estimation of Mean and Variance 24
2.4.2 Allocation of Sample Size 24
2.4.3 Advantages of Stratified Random Sampling 25
2.5 Cluster Sampling 27
2.5.1 Estimation of Population Mean 28
2.6 Multistage Sampling 30
2.7 Sample Size 31
2.8 Margin of Error 31
2.9 Fish Sampling 33
2.10 Water Sampling and Monitoring 33
2.11 Ecological Sampling Methods 33

2.11.1 Some Spatial Sampling Methods . . . 34
2.11.2 Density and Dispersion Patterns . . . 36
2.11.3 Determining the Dispersion Pattern . . . 36
2.12 Constraints of Data Collection in Inland Open Waters . . . 37
2.13 Conclusion . . . 38
Exercise . . . 38
Further Reading . . . 40

3 Descriptive Statistics for Inland Fisheries Data . . . 41
3.1 Population and Sample . . . 42
3.2 Parameter and Statistic . . . 42
3.3 Types of Data . . . 43
3.4 Data Quality . . . 43
3.5 Tabular and Graphical Presentation of Data . . . 44
3.5.1 Frequency Distribution Table . . . 44
3.5.2 Scatter Diagram . . . 44
3.5.3 Bar Diagrams . . . 45
3.5.4 Pie Diagrams . . . 46
3.5.5 Histogram . . . 47
3.5.6 Boxplot . . . 48
3.6 Measures of Central Tendency . . . 49
3.6.1 Arithmetic Mean . . . 49
3.6.2 Median . . . 49
3.6.3 Mode . . . 50
3.6.4 Geometric Mean (GM) . . . 50
3.6.5 Harmonic Mean . . . 50
3.6.6 Weighted Mean . . . 51
3.7 Measures of Dispersion or Variation . . . 51
3.7.1 Range . . . 51
3.7.2 Variance and Standard Deviation (SD) . . . 52
3.7.3 Standard Error . . . 52
3.7.4 Coefficient of Variation (CV) . . . 53
3.7.5 Shape of Data Set/Distribution . . . 53
3.7.6 Types of Distribution . . . 54
3.8 Conclusion . . . 55
Exercise . . . 55
Further Reading . . . 58

4 Basic Concept of Hypothesis Testing and Parametric Test . . . 59
4.1 Significance Level . . . 60
4.2 Type of Errors . . . 62
4.3 Power of Test . . . 62
4.4 Confidence Level, Limits and Interval . . . 63
4.4.1 Interpretation of Confidence Interval . . . 64
4.5 Selection of Statistical Tools . . . 66

4.5.1 One Sample Parametric Test 67
4.5.2 Large Sample Test of Significance for Population Parameter 67
4.5.3 Parametric Test of Significance Based on Two Samples 71
4.5.4 *t-Test* for Testing of Difference Between Two Means of Uncorrelated Observations 74
4.6 F-Test for Equality of Two Population Variances 76
4.7 Conclusion 79
Exercise 79
Further Reading 80

5 Chi-Square Test of Significance 81
5.1 Chi-Square Test for Goodness of Fit (One Sample Chi-Square) 82
5.2 Chi-Square Test for Independence 84
5.3 Measure of Association 90
5.4 Conclusion 90
Exercise 91
Further Reading 94

6 Use of Non-parametric Test in Fisheries Data 95
6.1 Concept of Non-parametric Test 95
6.1.1 Advantages of Non-parametric Hypothesis Test 97
6.1.2 Disadvantages of Non-parametric Hypothesis Test 98
6.2 Data Diagnostics for Using Non-parametric Test 98
6.3 Ranking of Data 99
6.3.1 Ranking with Ties 100
6.4 A Systematic Guide to Apply Non-parametric Test Procedures 101
6.5 Some Frequently Used Non-parametric Test Methods 103
6.5.1 Single Sample Problem 103
6.5.2 Comparing Two Dependent Samples 106
6.5.3 Comparing Two Independent Samples: The Mann-Whitney *U*-Test 108
6.5.4 Comparing Three or More Independent Samples: The Kruskal– Wallis Test 110
6.6 Conclusion 113
Exercise 114
Further Reading 117

7 Analysis of Variance (ANOVA) and Design of Experiments 119
7.1 Assumptions of ANOVA 120
7.2 Hypotheses and Procedure of ANOVA 120
7.3 One-Way ANOVA 121
7.4 Two-Way ANOVA 123
7.5 Repeated Measures ANOVA (RMANOVA) 128

7.6 Analysis of Variance and Experimental Designs 129
7.7 Experimental Designs 129
7.7.1 Principles of Experimental Design 131
7.8 Completely Randomized Design (CRD) 131
7.9 Randomized Block Design (RBD) 132
7.10 Conclusion 133
Exercise 133
Further Reading 136

8 Concept of Simple Correlation and Regression 137
8.1 Simple Correlation Coefficient 138
8.1.1 Some Properties of Correlation Coefficient 138
8.1.2 Distinction Between r and r^2 140
8.2 Test of Significance of Correlation Coefficient 140
8.3 Non-parametric Correlation or Rank Correlation 142
8.4 Concept of Regression Analysis 144
8.4.1 Properties of Regression Coefficient 147
8.4.2 Testing of Regression Parameters 147
8.4.3 Model Accuracy Measurement and Outlier Detection 148
8.5 Conclusion 157
Exercise 158
Further Reading 160

9 Analysis and Interpretation of Weight-Length Data of Fish 161
9.1 Concept of Weight-Length Relationship 162
9.2 Interpretation of Parameters 163
9.3 Condition Factor 163
9.4 Determining WLR from Observed Data 164
9.4.1 Analysis of Weight-Length Data Using R Software . . . 165
9.5 Determine Fulton's and Relative Condition Factor 166
9.5.1 Plotting Condition Factors Against Length 169
9.6 Conclusion 169
Exercise 170
Further Reading 171

10 Some Multivariate Analysis Techniques for Fisheries Data 173
10.1 Principal Component Analysis 174
10.1.1 PCA Workflow 175
10.1.2 Real Data Example 175
10.2 Multiple Regression Analysis 183
10.2.1 Working Principle 183
10.2.2 Workflow of Multiple Regression Analysis 185
10.2.3 Real Data Example 185
10.3 Conclusion 188
Exercise 189
Further Reading 193

11 Introduction to Electronic Spreadsheet and Microsoft Excel 195
11.1 What Is a Spreadsheet? 195
11.2 Spreadsheet Development Timeline 196
11.2.1 Beginnings and the 'Tale of VisiCalc' 196
11.2.2 Daniel Fylstra 196
11.2.3 After VisiCalc? 197
11.2.4 Lotus 1-2-3 197
11.2.5 Microsoft Excel 197
11.3 Other Electronic Spreadsheets 198
11.3.1 Google Docs 198
11.3.2 OpenOffice 198
11.3.3 Microsoft Excel 198
11.3.4 Excel Web Application 198
11.3.5 Zoho Sheet 199
11.4 Electronic Spreadsheets and Microsoft Excel 199
11.4.1 Uses of an Electronic Spreadsheet 199
11.4.2 Advantages of Using an Electronic Spreadsheet 200
11.4.3 Anatomy of an Excel Program Window 201
11.5 Functions Related to Inland Fisheries Management 203
11.5.1 Data Validation 203
11.6 Data Entry by Creating Forms 203
11.7 Conclusion 206
Exercise 206
Further Reading 208

12 Introduction to R Software 209
12.1 Why Use R? 210
12.1.1 Programming Features of R 210
12.2 Getting Started 210
12.2.1 Installation (for Windows Users) 210
12.2.2 Working with R Scripts 211
12.2.3 R Graphical User Interface (RGui) 211
12.2.4 RStudio 212
12.2.5 R Scripting 213
12.2.6 Sourcing a Script in R 213
12.3 Arithmetic and Logical Operators 214
12.3.1 Arithmetic Operators 214
12.3.2 Comments 214
12.3.3 Logical Operators 215
12.4 Data Types and Objects in R 215
12.4.1 Scalars 215
12.4.2 Vectors 217
12.4.3 Matrix 217
12.4.4 Data Frames 218
12.4.5 Lists 219
12.4.6 Objects 220

12.5 Handling Data Frames 221
12.5.1 Creating a Data Frame 221
12.5.2 Summary of Data 222
12.5.3 Extracting Data from a Data Frame 223
12.5.4 Expanding the Data Frame 223
12.6 Importing Data from Files 224
12.6.1 Working Directory 224
12.6.2 Importing from Excel 225
12.6.3 Importing from Table File 225
12.6.4 Importing from CSV Files 226
12.7 Conclusions 227
Exercise 228
Further Reading 233

13 Introduction to Graphics in R 235
13.1 Standard Graphics in R 235
13.1.1 Titles, Legends and Annotations Titles 236
13.2 Statistical Graphics Gallery 243
13.2.1 Line Plot 244
13.2.2 Scatter Plot 244
13.2.3 Scatter Plot Smoothing 244
13.2.4 Box Plots 245
13.2.5 Histogram Plot 246
13.2.6 Histogram with Density Estimate 247
13.2.7 Normal Q-Q Plot 248
13.2.8 Empirical Distribution Function Plot 249
13.2.9 Bivariate Contour Plot 250
13.3 Conclusion 252
Exercise 252
Further Reading 254

Appendix: Answers of All Exercises 255

About the Authors

Basant Kumar Das is currently the Director of ICAR-Central Inland Fisheries Research Institute. He has pursued his Ph.D. in Aquaculture and Postdoc in Molecular Immunology from Marine Lab, University of Aberdeen, Scotland. His subject of specialization is Aquaculture, Molecular Immunology, Fisheries, Nanotechnology and Fish Health. He has 26 years of professional experience in disease diagnosis of fish, diversification of Inland Cage, aquaculture practices and feed development. Dr. B.K. Das has developed Reservoir Fisheries Database and Fish Disease Advisory App in Inland Fisheries and also commercialized four technologies. He has authored 200 research papers, with more than 4000 citations, 30 books and numerous other souvenirs, book chapters and popular articles. He is the recipient of more than ten awards/honours, such as Dr. M.-S. Swaminathan Award for Best Indian Fisheries Scientist, Krushakbandhu and Krushak-Ratna by Krushkasamaj, Eminent Zoologist of the year Award from Zoological Society of India, Bodhgaya Fellow of International Society for Environmental Protection (ISEP), Fellow of National Academy of Agricultural Sciences (NAAS), DBT Overseas Associateship, Lal Bahadur Shastri Young Scientist Award (ICAR), Dr. Hiralal Chaudhuri Annual Awards and a Life Member of Indian Science Congress Association (ISCA) and Member of Executive Committee, ISCA (2020–2021) as well. He has also initiated projects on Anti-microbial Resistance (AMR) in Fisheries, Ranching of Hilsa above Farakka Barrage and Fisheries Development in Kothia Maun of Bihar.

Dharm Nath Jha is a senior scientist of Agricultural Statistics at ICAR-Central Inland Fisheries Research Institute, Prayagraj, with a sound knowledge and research background in statistics. He has a high level of expertise in data analysis viz. time series analysis, spatial analysis, regression analysis, deep learning, multivariate analysis along with descriptive statistics of data related to the fishery domain. As a Co-PI of the project he has completed five projects of ICAR-CIFRI and two externally funded projects. During these projects he worked on important issues related to aquatic ecosystem, fish and fishery. He estimated environmental flow requirement of Sone river, i.e. minimum amount of water to maintain the river ecosystem functional and intact. He has estimated the impact of river linking on ecosystem and fish and fishery of river Ken and Betwa. Fish catch and water resource estimation of Haryana district was completed by Him. He used the statistical tools and technique viz. time series analysis, spatial analysis, regression analysis, deep learning and multivariate analysis for the analysis and interpretation of data related to aquatic ecosystem. He has calculated economics of pen culture of different wetlands selected under these projects along with its analysis. The different types of statistical software known to him are SAS, SPSS, R, MATLAB, MS Excel, etc. He is a recipient of young scientist award in the year 2016.

Sanjeev Kumar Sahu works as a scientist in ICAR-Central Inland Fisheries Research Institute, Barrackpore, for the past 21 years. He has completed his graduation in electronics from post-graduation in control and instrumentation. His area of interest is the Application of Geospatial Technology, Geostatistics, database management, and online GIS in Natural Resource Management and fisheries research. He has published more than 15 research papers in international and national journals, five books (edited) and several book chapters. He has delineated surface water bodies of 19 states for fisheries purposes and developed e-atlas for waterbodies.

Anil Kumar Yadav works as scientist in the discipline of Agricultural Statistics at ICAR-Central Inland Fisheries Research Institute, Regional Centre, Guwahati. He joined Agricultural Research Service of ICAR in 2009 and has over 10 years of research experience in the field of data science for the management of inland open waters. He completed his M.Phil. in Statistics with specialization in data mining from the University of Rajasthan, Jaipur. He has published over 30 research papers in reputed national and international journals along with a number of popular articles and manuals. He has worked on a number of research projects of national importance.

Rohan Kumar Raman Scientist (Agricultural Statistics), presently works at ICAR Research Complex for Eastern Region, Patna. Dr. Raman has completed his graduation (2008) from Dr. Rajendra Prasad Central Agricultural University, Pusa (Samastipur), Bihar, Post-graduation (2010) and Doctor of Philosophy (2014) from Indian Agricultural Statistical Research Institute, Pusa, New Delhi. He has more than 7 years of professional experience in the field of data analytics and research. He is a two times recipient of gold medal young scientist awards (2017 and 2018) and one time best oral presentation awards (2014) from professional societies. He has developed calibrated estimators for survey sampling in non-response situations, time series forecasting model for total fish landing in Chilika lake, Odisha, and multispecies time series modelling tools for commercial fisheries in a tropical river-estuary system, India. He has also contributed as data analytical experts in the field of bioinformatics, crop sciences, veterinary and fisheries sciences. He has more than 50 national and international publications, book chapters and popular articles. He has experience of working on several research projects of national and international importance in the area of agricultural statistics.

M. Karthikeyan Scientist (SG), from Coimbatore in Tamil Nadu, received his bachelor's degree from Madras University and pursued his master's in 'Agricultural Statistics' from Indian Agricultural Research Institute (IARI), New Delhi. Subsequently, he got selected to Agricultural Research Service (ARS) and joined as Scientist at Central Marine Fisheries Research Institute (ICAR-CMFRI) at Kochi, Kerala, in the year 1986 where he worked for 4 years. In 1990, he joined the Regional Centre of Central Inland Fisheries Research Institute (ICAR-CIFRI), Bangalore, and is continuing here since then.

His main areas of research have been development of sampling methodologies for the estimation of fish catches, development of software for fish catch estimation, development of statistical models for fish yield prediction and to determine sustainable fish harvest levels, etc. His significant contribution in the recent past has been the development of '*eMatsya*'—an innovative electronic data acquisition system (mobile-based Android App) to collect fish catch data from inland water bodies. He received DAAD Fellowship from the Government of Germany for his master's programme 'M.Agrar' from the University of Bonn, Germany, during 1994–1996. As one of the Indian delegates, he visited the Food and Agriculture Organization (FAO), Rome, Italy, in 2009 for discussions on various aspects of 'Fishery and Aquaculture Statistics' with FAO experts. He strongly believes that moving forward, all cutting-edge technologies emerging in various fields should be prudently harnessed in fisheries, especially in inland fisheries, to bring about the 'Blue Revolution' in India.

List of Figures

Fig. 1.1 A data-driven model for decision-making 2
Fig. 1.2 A typical data structure to study species environment interaction 10
Fig. 1.3 Basic analytical framework for species environment data 11
Fig. 1.4 A spatial data presentation on Google Earth 12
Fig. 2.1 Schematic diagram of a three-stage sampling 30
Fig. 2.2 Margin error and sample size relationship 31
Fig. 2.3 Selection of sampling site using line transect method 35
Fig. 3.1 Number of boats operating and total catch per day 45
Fig. 3.2 Fish preference by consumers in Tripura 45
Fig. 3.3 Multiple bar diagram of fish production in unstocked, stocked and all beels of Assam 46
Fig. 3.4 Component bar diagram of fish production of stocked and unstocked beels in Assam 47
Fig. 3.5 Pie diagram for fish group composition by weight 47
Fig. 3.6 Histogram, frequency polygon and frequency curve for fish length data set 48
Fig. 3.7 Boxplot for fish yield in stocked and unstocked beels 48
Fig. 3.8 Different types of skewed distributions 53
Fig. 4.1 Normal plot with 95% confidence limit between $-1.96\ \sigma_{\bar{x}}$ and $1.96\ \sigma_{\bar{x}}$ 64
Fig. 4.2 A generic $(1-\alpha) \times 100\%$ confidence interval for the mean of a normally distributed population 65
Fig. 4.3 Screenshot results of paired *t-test* in SPSS package 73
Fig. 4.4 Screenshot of paired *t-test* in SPSS package 74
Fig. 4.5 Screenshot of paired *t-test* in SPSS package 74
Fig. 4.6 Screenshot of two sample *t-test* in SPSS package 76
Fig. 4.7 Screenshot for F-test in excel 78
Fig. 6.1 Fish yield distribution of two districts 99
Fig. 6.2 Yield distribution in different management categories 113
Fig. 7.1 Schematic data format for different ANOVA 132

Fig. 8.1 $r = 1$, perfect positive correlation, and $r = -1$, perfect negative correlation ... 139
Fig. 8.2 Strong correlation: (**a**) positive and (**b**) negative ... 139
Fig. 8.3 Weak correlation, (**a**) positive and (**b**) negative ... 139
Fig. 8.4 Almost zero correlation ... 139
Fig. 8.5 Almost zero correlation with no linear relationship but strong nonlinear relationship ... 140
Fig. 8.6 The plots showed the fish production per year increases as the nitrogen level increases. (**Source:** Bhujel 2008) ... 145
Fig. 8.7 Slope of the line as relationship between factor X (nitrogen fertilizer level) and response variable (fish production). (**Source:** Bhujel, 2008) ... 145
Fig. 8.8 Relationship between factor and variable which may be either positive (+), negative (−) or (0) zero depending on the situation. (**Source:** Bhujel 2008) ... 146
Fig. 8.9 The scatter plot of fish weight and age (weeks) ... 150
Fig. 8.10 Residual of weight vs weeks plot after fitting the model ... 151
Fig. 8.11 Fit plot of fish weight against weeks ... 151
Fig. 8.12 Regression model fit diagnostic plot for fish weight gain at different age (weeks) ... 152
Fig. 8.13 Plot of fish weight versus age ... 153
Fig. 8.14 Residual for fish weight plot vs weeks after fitting linear model ... 153
Fig. 8.15 Fit diagnostic plot after fitting linear model ... 154
Fig. 8.16 Regression fit plot after linear model fitting on fish weight on age ... 155
Fig. 8.17 Residual for fish weight plot vs weeks after fitting linear model ... 156
Fig. 8.18 Best fit line after changing the outlier and fit the regression line ... 156
Fig. 9.1 The weight-length relationship of the studied fish species ... 167
Fig. 9.2 Fulton's and relative condition factor for different months ... 168
Fig. 9.3 Fulton's and relative condition factor for different places ... 168
Fig. 9.4 Fulton's and relative condition factor against length ... 169
Fig. 10.1 Variance explained against the number of principal components ... 177
Fig. 10.2 Structural dependence of variables with principal component ... 180
Fig. 10.3 Score distribution pattern of 14 rivers against two main components ... 181
Fig. 10.4 Distribution of errors for non-transformed species richness ... 187
Fig. 10.5 Distribution of errors for log-transformed species richness ... 187
Fig. 10.6 Relative Importance of predictors of fish species richness ... 189
Fig. 13.1 Setting margin and margin text in R graphics ... 238
Fig. 13.2 Types of graph produced by plot function, for different values of type argument ... 239

Fig. 13.3 Axis style change using 'las' argument 241
Fig. 13.4 The graph with two y-axes, representing catch and effort, respectively 241
Fig. 13.5 A fitted curve added to the scatter plot 243
Fig. 13.6 Scatterplot smoothing to explore relationship 245
Fig. 13.7 Histogram overlaid by normal density 247
Fig. 13.8 Histogram with overlaid kernel density estimate 248
Fig. 13.9 Q-Q plot to examine the normality. This example is for the weight distribution of Hilsa 249
Fig. 13.10 Empirical cumulative distribution function for the weight distribution of Hilsa 250
Fig. 13.11 Contour plot for identification of breeding ground of Hilsa 251

List of Tables

Table 1.1	Data structure for stocking structure based on physical habitat, water quality and other biotic community	13
Table 1.2	Variable, their coding and measurement unit of surveyed data	14
Table 2.1	Fish landings of 10 randomly selected days	21
Table 2.2	Fish landings for 10 days	23
Table 2.3	Monthly fish catch (in kg) for sampled clusters	29
Table 2.4	z^* values at different confidence levels	32
Table 3.1	Parameter and statistic	42
Table 3.2	Frequency distribution of fish length data	44
Table 3.3	Observations on the boats operating and daily catch in the river stretch	45
Table 3.4	Consumers' preference for different fish species in Tripura	45
Table 3.5	Fish production (MT) in beels under AFDC from 2010 to 2016	46
Table 3.6	Threshold values for categorizing skewness and kurtosis of a distribution	54
Table 4.1	Critical values for Z-test at 0.05 and 0.01 significance level	61
Table 4.2	Types of errors in hypothesis testing	62
Table 4.3	The critical values (Z_α) for Z	68
Table 4.4	DO value and fish weight of sampled fish from the pond	70
Table 4.5	Paired samples statistics	72
Table 4.6	Screenshot of two sample *t-test* data in MS Excel	77
Table 4.7	Screenshot of SPSS output	77
Table 4.8	Screenshot of SPSS output	77
Table 5.1	Observed frequencies for different fish species in Tripura	83
Table 5.2	Output of chi-square analysis in SPSS	90
Table 6.1	Fish species richness in river 1 and river 2	100
Table 6.2	Fish species richness in river 3 and river 4	100
Table 6.3	Ranking fish species richness with ties	101
Table 6.4	Most popular non-parametric tests and their parametric equivalent for different analyses	101

Table 6.5 Computation sheet for Wilcoxon T ... 105
Table 6.6 Parasite infection in two body sectors of fish ... 107
Table 6.7 Computation of test statistics for paired sample test. S_1: sector 1; S_2: sector 2 ... 108
Table 6.8 Fish yield of 23 beels in two districts ... 109
Table 6.9 Ranks of yield of surveyed wetland for different management categories ... 112
Table 7.1 A typical format of one-way ANOVA ... 121
Table 7.2 Fish yield for different levels of stocking ... 121
Table 7.3 ANOVA table for effect of factor stocking on yield of beels ... 123
Table 7.4 Duncan post hoc analysis. Subset for alpha =0.05 ... 123
Table 7.5 A typical format of one-way ANOVA ... 124
Table 7.6 Fish yield for different stocking status and macrophyte infestation ... 124
Table 7.7 Results of two-way ANOVA ... 126
Table 7.8 Different design of experiment and their analytical methods ... 130
Table 8.1 Length and weight of the *S. strongylura* collected from a lake and their square value (L^2 & Wt^2) and product value ($L \times Wt$) ... 141
Table 8.2 Egg viability and ammonia nitrogen level ... 143
Table 8.3 Weights of fish (gm) from 0 to 12th week of age ... 148
Table 8.4 Model parameter estimate and their significance and fit statistics using linear regression model fitted ... 149
Table 8.5 New weight gain of the fish at different age (weeks) ... 152
Table 8.6 Model parameters estimate and their significance and fit statistics using linear regression model fitted ... 155
Table 8.7 Parameter estimates and their statistical significance ... 155
Table 8.8 Model parameters estimate and their significance and fit statistics using linear regression model fitted ... 157
Table 8.9 Parameter estimates and their statistical significance ... 157
Table 9.1 An abridged dataset of weight-length of a fish species ... 165
Table 10.1 Fish species richness and climato-hydrological variables of 14 major rivers of India ... 176
Table 10.2 Pairwise correlation among the climato-hydrological variables. (1) Temperature range; (2) Rainfall; (3) Discharge; (4) Sediment; (5) Basin area; (6) Latitude; (7) Water temperature ... 177
Table 10.3 Loadings of each variables to the extracted factor ... 179
Table 10.4 Results of multiple regression ... 188

Nature and Sources of Data for Inland Fisheries Management

1

Key Topics

- Data-driven fisheries management
- Data requirement
- Management complexities
- About the routinely collected data

A couple of decades ago, strategic planning of inland fisheries management was mostly 'knowledge-driven', entailing information based on published literature, experiences in the field survey, flow of knowledge from fishermen to the researcher or policymaker, etc. After acquiring that knowledge, researchers would normally apply their intelligence to recommend management strategies. However, data and information on both the users and resources of inland fisheries are regarded as essential for use in long-term planning and policy analysis and aquatic resource-specific strategic management. Very recently, there has been a paradigm shift from 'knowledge-driven' to 'data-driven' management of inland fisheries in India. The 'data-driven' fisheries management emphasizes on the quantitative data, analysis and meaningful interpretation. The results emanated by analysing quantitative data further generate valuable information that enhances the knowledge of the researchers. These data and generated information consequently become helpful to formulate effective management strategies for inland fisheries so that the general public, researchers, trained managers and administrators can easily understand and see clearly the trade-off involved among all possible alternatives. A typical resource-specific 'data-driven' decision-making process is conceptualized, as shown in Fig. 1.1. However, the flow of information and accumulation of knowledge are not the only model for fisheries management. It depends on the way of gathering information and presentation by the researchers. To know what kind of data and

B. K. Das et al., *Concept Building in Fisheries Data Analysis*,
https://doi.org/10.1007/978-981-19-4411-6_1

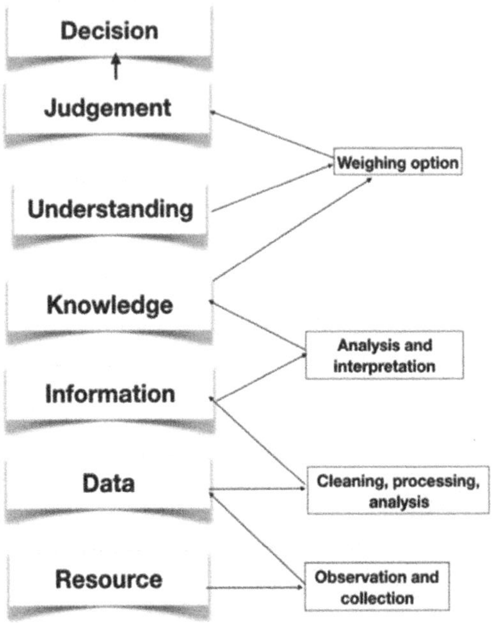

Fig. 1.1 A data-driven model for decision-making

information is needed for inland fisheries management, one has to monitor and analyse fisheries management as a continuous decision-making process.

1.1 Complexity of Inland Fisheries Management

Management of inland and marine fisheries distinctly differs, although some common definitions and concepts are used in both environments. There are numbers of characteristics, as mentioned below, which distinguish inland fisheries from marine environment; thereby the fisheries management applied to them.

1. There is great physical as well as ecological diversity among the inland water ecosystem, e.g. rivers, lake, pond, tanks and small streams. Particularly in India, huge physical diversity exists due to geomorphologic variations of the country. For example, characteristics of rivers and associated fisheries in the Himalayan region are entirely different from the plain lands. So, common management strategies for all the rivers of India will not be effective.
2. Usually the order of freshwater ecosystem is much smaller than that of marine ones. The number of water bodies, on the other hand, is much larger in inland environment. Moreover, there is huge diversity in the nature and magnitude of exploitation of inland open water. This indicates that intensive management cannot cover all inland water bodies equally. Thus a selection, either for representative water bodies or extensive management, has to be made.

3. In some cases, maintaining water quality is of prime importance. So, formulation of management of strategies becomes very difficult, which needs separate assessment process.
4. Multiple uses of water in inland system further complicate strategic resource-specific management.

1.2 Data Requirement

Any aquatic resource can be managed efficiently and effectively when the right kind of data and information is available in both quantity and quality. In most of the countries, recording and collection of data in inland fisheries are essentially weak and, in general, insufficient for prudent 'data-driven' decision-making. It is, therefore, need of the hour to collect and preserve better data and information on the inland water environment, the fish and the fishers that form essential components of inland fisheries. In the previous section, vastness and complexity of inland fisheries management have been described. It can be contemplated that requirement of data for inland fisheries management is diverse in nature.

1.2.1 Objectives

The collection of data is not an end in itself, but is essential for informed decision--making. Fishery is a complex system of interacting factors incorporating the biological resource, social and institutional constraints, economic conditions and cultural beliefs. The greatest threats to these fisheries are often not overexploitation of resources, but rather loss and degradation of the aquatic habitat and poor land use practices that lead to sedimentation and pollution. Under these circumstances, the management of fish habitats and related environments is often a priority, so the collection of environmental and ecological data in relation to fishing patterns is essential. Integrated analysis, using a variety of data, is necessary for predictive assessments of future conditions and the outcomes of alternative management measures.

Planners and managers need to understand the dynamics of the fish stocks, fishery operations, infrastructure, communities and individuals involved in the fisheries sector in order to set policy and manage fisheries. Data collection and analysis, for example, can provide information on how fishers are likely to respond to different policies. Constraints on production and development of new fisheries can be identified. Prices and cost changes in the fisheries can be assessed. Stocks likely to receive increased levels of exploitation may be identified before resource levels drop to a crisis point. As ecosystems worldwide have become subject to increasing stress, it has become more important to develop, maintain and improve harvesting regimes which minimize negative impacts on habitats and fish communities. Specific data on fishing operations, fishers, fishing communities and the environment are required to achieve this objective.

Specifically, the nature of data required highly depends on aquatic resource type as well as management objectives that can be broadly categorized as follows:

- Extractive objectives, e.g. recreational opportunities or fish protein supply
- Social objectives, e.g. income generation, rationalized distribution of benefits among stakeholders, reduction of social conflicts, etc.
- Fiscal objectives, e.g. revenue and foreign earnings
- Conservation objectives, e.g. sustainable fisheries or biodiversity conservation or combination of both

Most of the inland fisheries of India are managed by setting mix of the aforementioned objectives. It is of significant importance that common approach of data collection mechanism must be devised so that it can serve most of the stakeholders involved in the fisheries.

1.2.2 Data Type

Before the beginning of data collection, it should be clearly defined which type of data one wants to collect. It may be operational data, biological data, economic data or socio-cultural data. Each data type may be used for a variety of indicators. Catch, for instance, may be used both in calculations of revenue for economic purposes and as a rough measure of resource depletion.

The selection of a data type also depends on the available analyses. Many fish populations' dynamics models require catch in weight and number by species, as well as other data on the biology of each species (e.g. age). A bioeconomic model may require data not only on the specific fishery production and prices but also on other economic sectors for comparative purposes.

For socio-cultural data, the essential starting point is data on individual fishing. Fish dealers and processors are the next most critical group. Data collection on other interested parties (such as consumers, environmental organizations, coastal developers, etc.) can be added based on the available cost. Some data may be acquired from existing routine data collections, such as fishing licences or permits and census records. Other data may need to be collected through new programmes.

1.2.3 Data and Information

It is pertinent to distinguish between data and information. Data are raw numerical values that must be analysed to generate information. Information, on the other hand, may not necessarily be numerical. It can be descriptive. For example, information may deal with a question such as whether a particular fish species migrates or not. Thus data and information together form the basis for fisheries management. The nature of data can be of two types:

- Absolute—it attempts to assign values to variables (e.g. catch) that are as close to reality as possible and may be recorded over space and time.
- Relative—it attempts to assign values to trace the movement of values over space and time with respect to a baseline.

There are varieties of subjects on which data and information may be gathered for formulating management strategies. These include the following:

- The fish—species and their abundance, feeding, breeding, growth, mortality, migration and other aspect of biology
- The fishery— catch statistics, effort, information on gear and its performances
- The fishers—socio-economics of fishers and their families
- The environment—including water quality, quantity, degree of modification of aquatic habitat and ecosystem

1.2.3.1 Some Indicators and Associated Data Types

There are many possible data types beyond those discussed here. However, the examples given should cover the most important ones.

Catch Catch in numbers or weight represents the removal of biomass and individuals from the ecosystem. Catches should be broken down into categories. A detailed breakdown can also improve economic and socio-cultural analyses. In most cases, it is useful to obtain catch both in weight and numbers. Conversion from numbers to weight (or vice versa) can be obtained through an estimate of the mean weight of individual fish caught. Length measurements may also be converted to total weight of the catch, if a reliable length–weight relationship has been established. Where discarding takes place, catches will not be the same as the live weight equivalent of the landings. Discarding has significant biological implications and should always be recorded or estimated. Total catch consists of total landings and discards.

Effort Effort in biological assessments is used to estimate fishing mortality. Fishing mortality is a fundamental variable in stock assessment, representing the proportion of the stock that is removed by fishing. In economic and socio-cultural analyses, effort can be related to fishing activity, vessel and fleet profitability and economic efficiency. To relate effort to fishing mortality for use in biological models, it is necessary to relate it very closely to specific gear use, such as trap soak time or time trawling. On the other hand, to relate effort to profitability requires data at the trip level, including time spent at sea, time fishing and labour and capital inputs.

Catch Per Unit Effort (CPUE) CPUE or catch rate is frequently the single most useful index for long-term monitoring of the fishery. It is often used as an index of stock abundance, where some relationship is assumed between the index and the stock size.

Fishing Operations Fishing operation indicators describe the composition of fishing fleets and fishing patterns and are the basis of most management decisions. They are important for monitoring compliance and in analyses involving fishing effort. For instance, mapping fleet activities by gear use allows management to detect infringements of zone allocations or potential conflicts in gear use (e.g. trawling versus gill net) which require zoning. Linking fishing operations to socio-cultural, infrastructure and other economic data improve analyses of fleet activities. Such analyses produce a better understanding of motivations in the behaviour of different fleets, so more accurate predictions can be made of the fleets' response to changes in the fishery.

Stock Enhancement Capture fisheries, particularly of inland waters, are increasingly being subjected to practices designed to increase the stock size and productivity of the fishery. Such practices include releasing juvenile stages produced in hatcheries into the wild habitat (called ranching), the introduction of fertilizer to ponds, lakes and reservoirs, the culling of predator species and the construction of artificial reefs to provide habitat for certain exploited species. Typical indicators are the fish production, the number of fish by species and age at introduction and subsequently; nutrient level, the number of fish before and after fertilizer use; and cost, the cost for research and development, implementing and monitoring.

Economic Indicators A number of measures exist that have been used by various agencies to measure the economic significance of fisheries to the national and local economies and assess performance of fisheries management in achieving economic objectives.

1. Market prices at the various market levels are short- and medium-term indicators of the demand for fish products. They signal changes in markets and, if properly interpreted, provide hints to the future commercial operations of the sector. Prices are also necessary for the calculation of many other economic indicators.
2. Gross value of production (GVP) and of processed products: The gross value of production (GVP) is determined by multiplying total production by the price received. GVP provides an indication of the potential economic importance of a fishery relative to other fisheries or other industries in a nation or province.
3. Costs and earnings.
4. Investment.
5. Management costs.
6. Gross value added: Gross value added (GVA) is the total amount paid as returns or rent to labour and capital (and theoretically to the resource base as well, although this rarely occurs without property rights).
7. Subsidies
8. Resource rent and economic profits
9. Domestic food supply and fish consumption
10. Employment level in the fisheries sector
11. Balance of trade of fish and fishery products

Socio-cultural Indicators Socio-cultural indicators are critical for evaluating policies and management activities, as they measure the value of fisheries beyond their simple economic worth. However, routine data collection for socio-cultural indicators has often been neglected, reliance being placed on ad hoc political procedures which aim to represent socio-cultural views. In practice, these will not replace objective assessments of performance, and socio-cultural data should be collected routinely alongside biological and economic information. These indicators usually include distribution of fishing income; distribution of fish consumption; nature of access to the fishery; demographics and fishing patterns in the harvesting sector; demographics and employment patterns in the processing and marketing sectors; and social status of fishing, e.g. the social status of fishers and perceptions of fishing as an occupation influence the likelihood of entry into and exit from the fishery.

1.2.4 Location and Time

Fisheries management is a continuous process. Location is also very important for local scale management. Data must be recorded over geographical space and time so that precise location-specific management strategies can be formulated. More specifically, every data point must be stamped by geographical location and time.

1.2.4.1 Data Collection Frequency

Data should be collected depending on their rates of change and the costs of measurement.

Very frequent: Very frequent data are usually collected by automatic recorder machine. The volumes of data can become inordinately large, and some pre-processing is necessary before the data is stored. Depending on the use of the data, the frequency may be reduced to a single daily record.

Daily: Daily changing phenomenon/event is recorded daily, such as catch, effort, processing rates, etc.

Monthly: Measurements based on months are appropriate for variables that change slowly and those that have a seasonal pattern, e.g. monthly rainfall, retail price index, etc.

Annual: This is used for slow-moving variables, e.g. investment in gear and vessels, annual fish production, total spawn collection, etc.

Trip: Harvest-related data can be reported at the end of each trip, including landings, a trip summary of effort, fishing grounds, prices, trip costs and other operational and micro-economic data. Although many variables are naturally collected by trip, not all trips need to be covered, but a sampling strategy can be used to reduce costs.

Infrequent: Other types of data can be collected at periods greater than a year or data which may be updated in every 3–5 years.

Note *Data on habitat degradation, household and demographic information data, etc. are usually collected infrequently.*

1.3 Data Collection Methods

The choice of method is influenced by the data collection strategy, the type of variable, the accuracy required, the collection point and the skill of the enumerator. Links between a variable, its source and practical methods for its collection can help in choosing appropriate methods. The main data collection methods are as follows:

1. Registration: A register is a depository of information on fishing vessels, companies, gear, licenses or individual fishers. It can be used to obtain complete enumeration through a legal requirement. Data on vessel type, size, gear type, country of origin, fish holding capacity, number of fishers and engine horsepower should be made available for the registry.
2. Questionnaires: In contrast with interviews, where an enumerator poses questions directly, questionnaires refer to forms filled in by respondents alone. Questionnaires can be handed out or sent by mail and later collected or returned by stamped addressed envelope. This method can be adopted for the entire population or selected samples.
3. Interviews: In interviews information is obtained through inquiry and recorded by enumerators. Structured interviews are performed by using survey forms, whereas open interviews are notes taken while talking with respondents.
4. Direct observations

   Observers: Observers can make direct measurements on the fish catch at landing sites, processing plants or in markets. The variables that enumerators can collect include catch (landing and discards), effort, vessel/gears, operations, environmental variables (e.g. soil and water quality parameter, temperature), biological variables (e.g. length, weight, age), the values and quantities of landings and sales, plankton population and benthic population.

   Inspectors: Inspectors are a kind of enumerator involved in law enforcement and surveillance (for fishing regulations, sanitary inspections, labour control, etc.). They may work at sea on surveillance vessels, at landing sites on shore, at processing factories and at markets. In general, scientific data are better collected by enumerators who are not directly involved in law enforcement. Nevertheless, many variables collected by the inspectors are very useful and include landings, operational information, effort, landing price, processing procedure and values of product to the market and processors. Inspectors are also useful in collecting employment data.

   Scientific research: Ecological research methods can be undertaken independent of commercial fishing operations to measure variables related to fish populations or the environment. Such research can be carried out by institutional research vessels or by industry or institutions using commercial fishing vessels. The objective is to obtain observations on biological (e.g. stock abundance or spatial distribution and fish size, maturity and spawning activities) and environmental (e.g. soil and water quality) variables. It is important that this type of research is carried out periodically in order to obtain time sequential data. Similarly, socio-cultural research methods can be used to obtain specific information useful to management. Although these methods may not often be considered routine, they provide important data and should be considered for infrequent data collection where possible. The key informants are individuals with specialized knowledge on a particular topic.

Data logging: Automatic location communicators (ALC) automatically log data through positioning and communications technology. They allow remote observation through recording of fishing activities and could replace logbooks and observers/inspectors on the bridges of fishing vessels.

5. Reporting: In most complete enumeration approaches, fisheries staff do not directly undertake data collection but use external data sources. Most commonly, these sources are data forms completed by the fishing companies themselves, middle persons, market operators, processors and even trading companies and custom offices. Such methods are almost exclusively used for semi-industrial and industrial fisheries and institutions. Data submitted by companies are related to harvest data, post-harvest data, sale, trade, etc.

The choice of the many methods for collecting fishery data will depend on the variables to be measured, the source and the resources available. For the same variable, the methods can be different depending on the type of fishery. For example, for a large-scale fishery, catch data would be best collected from logbooks, while in a small-scale fishery, interviews and/or questionnaires would often be the best method. The sources (fishers, processors, etc.) are also an important factor for the choice and design of methods. Buyers, processors and other intermediaries are likely to keep their own sales records, which should be used as the basis of data forms. Small-scale fishers often do not keep any records, and data acquisition in this case would be restricted to one-to-one interviews, but the interview structure could be more flexible.

Many variables can be collected by more than one method and at different points from fishers to consumers. Where possible, data should be collected from several sources to cross-check for errors. In almost all cases, many different variables can be collected simultaneously. For example, length frequency, species composition, average weight and first sale price can all be obtained when vessels land their catch. Collecting of data for different purposes reduces costs, and thus due account should be made of this aspect when planning the data collection.

1.4 Some Frequently Collected Data in Inland Fisheries

This section describes some useful data elements essential for inland fisheries assessment. The data requirement specific to a particular assessment is delineated as follows:

- Stock assessment: Unlike marine environment, stock assessment is not so important for inland fisheries management. However stock assessment becomes essential for some of the water resources such as estuaries and large reservoir.

Structured data required for stock assessment demands huge cost and manpower. There are limitations for data collection in inland fisheries. Consequently the data is not so rich to perform stock assessment in inland water, especially in India. There are some basic data, which can be gathered easily and are useful for stock assessment in a data-poor condition.

- Catch-based stock assessment: The basic element for data-driven management of any fisheries is the catch data. The catch data is the final output by which the fisheries is monitored. So it is essential to record catch data for inland water resources. As far as stock assessment is concerned, catch-based method is the simplest one. Normally, surplus production model is used for assessing the stock. Catch and catch-per-unit effort are the two kinds of data required for such assessment.
- Length-based stock assessment: Commercial fishing is size-selective. It subjects to spatial and temporal variation in size of the exploited population. Collection of size data of fish is relatively easy. It can be utilized for length-based stock assessment. It is used to establish the growth pattern of any species. There are a number of stock assessment tools based on size. Interested readers may consult ICES (2015) for details of stock assessment tools, which rely on basic statistical tools.

1.4.1 Data for Biotic–Environment Assessment

Inland fisheries management is mostly based upon studies on biotic–environment interaction. More specifically, the issue addressed here is how biotic community is interacting with the aquatic environment. Conservation and sustainable fisheries are given main emphasis in such assessment. Typically the data are collected on varieties of subjects as shown pictorially in Fig. 1.2. Though this is very often used in riverine system, it can also be used in other inland open water. Usually the aquatic environment is divided into different homogenous strata according to

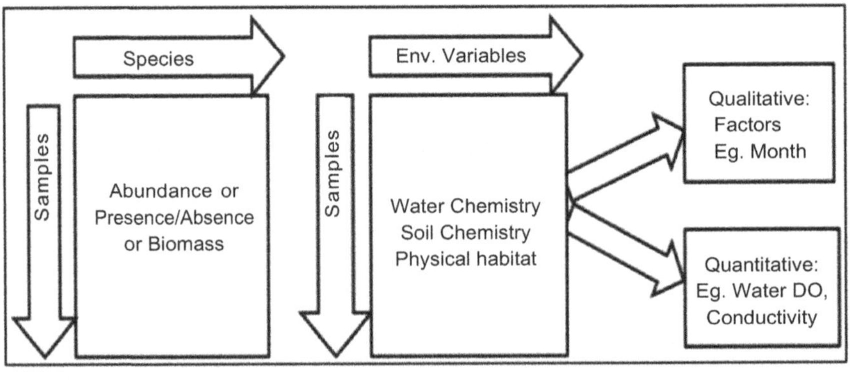

Fig. 1.2 A typical data structure to study species environment interaction

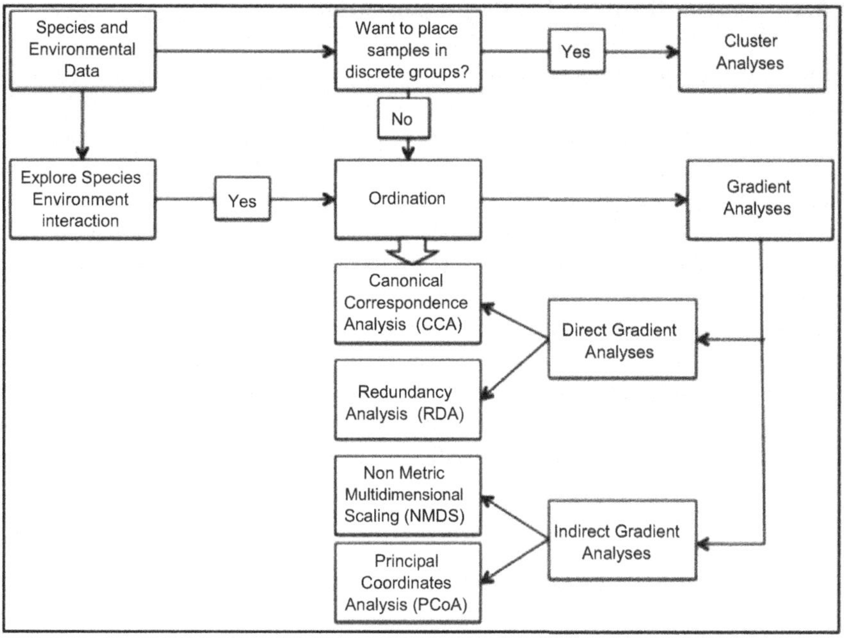

Fig. 1.3 Basic analytical framework for species environment data

ecological, physical habitat or fishery homogeneity. Thereafter, some sites are selected from each stratum, to accommodate spatial uncertainty. In each of the study sites, sampling is conducted at fixed time interval (weekly, monthly, bi-monthly). Sampling frequency depends upon the cost involvement, nature and objective of the study. Thus the resulting number of samples is the product of number of sampling sites and frequency over time. The species data can be the fish, the plankton, the benthos or any biotic community. The type of the data may be presence or absence. It may so happen that species occurrence data are the only present data. This very often occurs when occurrence data is recorded based on the interview of fisherman. Moreover, one can only ensure the presence if the species is encountered during sampling. It may be the case that species is present but it has not been detected. This is called pseudo absence data. While the analysis of presence/absence data is straightforward, presence only data requires analytical expertise.

The species abundance data is usually recorded either in count or in biomass. The plankton abundance is by default recorded as count. To generate useful information and subsequent interpretation, this kind of data is analysed. Application of suitable statistical tools for analysis depends on the objectives of the investigation. Although there is no thumb rule for any data-specific analysis, a basic framework is provided in Fig. 1.3. The details of carrying out the mentioned analysis are beyond the scope of this book. Interested reader may consult further (Borcard et al. 2011; McCune and Grace 2002; Braak and Ter 1986; and references therein).

1.4.2 Spatial Data for Fishery Assessment

To formulate localized management strategies, collection and recording of fishery data on spatial scale are essential. It will also help to characterize the fishery on spatial scale of desired resolution. Grid-based sampling can be adopted on particular aquatic resources. Administrative boundaries can also be the spatial resolution of the data. An example of spatial data is given in Fig. 1.4. The catch and effort data of Hilsa were collected in the Hooghly–Mathlah estuary region. Geo-referenced catch data was recorded by using navigation tools. Geographical locations (e.g. longitude and latitude) and times of the start and the end positions of each haul of gear operation were also recorded. These kinds of data are usually obtained by grid-based experimental fishing. The commercial fishing data can also be recorded by the commercial vessel equipped with navigation tools. This forms a good spatial data. This data can be utilized to identify overexploited zone so that fishing can be restricted to those zones. Location-specific management strategies can be formulated for the fisheries in the estuary.

1.4.3 Data for Stock Enhancement Strategies

Stock enhancement strategies are followed to enhance the fish yield of wetland and reservoir. Here the researchers attempt to devise optimal stocking strategy for sustainable fish yield of wetland or reservoir. To devise common stocking strategies for similar water bodies, data is usually recorded on several similar wetlands. To recommend optimal stocking based on the water quality parameter of the water

Fig. 1.4 A spatial data presentation on Google Earth

Table 1.1 Data structure for stocking structure based on physical habitat, water quality and other biotic community

Water body	Stocking		
Yield	Physical habitat	Water quality	Plankton/benthos
1			
2			
…			
…			
n			

body, the following data elements are recorded. The data structure is in Table 1.1. If the stocking strategies are to be recommended in a single water body, then the data on the same structure over time (repeated over several years) have to be recorded.

1.4.4 Survey Data

1.4.4.1 Example 1.1: Floodplain Wetland Fisheries

Usually fisheries management in wetlands focuses on sustainable fisheries so that yield can be optimized. If management regime already exists, it is necessary to evaluate regularly the performance. Essentially prevalent management regime is spread over spatial region. To evaluate this strategy, survey-based method is applied for data collection. Stratified sampling method is applied for the survey. Stratification is usually done according to administrative boundary, representing each district as stratum. The sampling frame in terms of district-wise distribution of floodplain wetlands is recorded. Structured questionnaire is designed to collect relevant information for assessment. The critical part of the survey is to design the questionnaire. First, the key attributes of measure of effectiveness of management are first identified. The above example is just a case study. This is a part of a broad socio-economic survey. The abridged version of data is shown in Table 1.2.

1.4.5 Perception-Based Data

When quantitative data is not available or scarce, perception-based study is normally adopted. Perception of the experts on the aquatic resources can be used for assessing the open water resources. Data and its preservation in inland fisheries of India are poor, albeit some efforts have been taken to strengthen it. In such situation data emanated from perception-based study can be of great use. The stakeholders involved in any exploited aquatic resources have wealth of knowledge regarding ecosystem status, changes in the aquatic ecosystem, as they are the integral part of the system. The knowledge and experience acquired by the stakeholders, fishers in particular, are extremely valuable for research and management of fisheries, especially in small-scale fisheries that are the major contributor to the inland fisheries. However, a major impediment to integrate stakeholders' knowledge in fisheries

Table 1.2 Variable, their coding and measurement unit of surveyed data

Sl. No.	Name of the explanatory variables	Variable type and unit	Broad category
1	Average area	In ha	Physical
2	Average depth	In meter	Physical
3	Connectivity	Close; open; semiopen	Physical
4	Stocking type	Autostocking; auto and artificial stocking; artificial stocking	Fisheries management
5	Pen culture	Absent or present	Fishery management
6	Pond culture	Absent or present	Fishery management
7	Weed regulation	Absent or present	Fishery management
8	Weed clearance	Absent or present	Fishery management
10	Gear restriction	Absent or present	Policy
11	Fishing restriction	Absent or present	Policy
12	Restriction on juvenile catch	Absent or present	Policy
13	Restriction of brooder catch	Absent or present	Policy
14	Lease holder	Individual; cooperative or open access	Policy
15	Yield	kg/ha/year	Key attribute to measure the effectiveness of management

management is the lack of formal guidance on using that knowledge. The description of the method of using stakeholders' knowledge is beyond the scope of this chapter. However, interested readers may consult Fischer et al. (2015) for ecosystem-based fisheries management, Mukherjee et al. (2015) for ecology and biological conservation and other available literature.

1.5 Conclusion

This chapter provides a brief overview of inland fisheries management scenarios and its complexities. Basic differences between 'knowledge-driven' and; 'data-driven' inland fisheries management are pointed out, and the role 'data-driven' approach is stressed for better strategic management. The chapter describes only the basic data requirement and their specific role for management. The data collection methods described are not exhaustive, and only the summarized version of different methods are described in the reviewed literature. The example data sets are mostly generated from the regular monitoring system of water resources. In fact, different inland open-water system requires different management protocols; thereby nature and data collection methods and their utilization differ substantially. It is highly recommended to consult the suggested reading to acquire advanced knowledge.

Further Reading

Bartley DM, De Graaf GJ, Valbo-Jørgensen J, Marmulla G (2015) Inland capture fisheries: status and data issues. Fish Manag Ecol 22(1):71–77. https://doi.org/10.1111/fme.12104

Borcard D, Gillet F, Legendre P (2011) Numerical ecology with R. Springer, New York

Braak C, Ter JF (1986) Canonical correspondence analysis: a new eigenvector technique for multivariate direct gradient analysis. Ecology 67(5):1167–1179

Diogo H, Pereira JG, Schmiing M (2017) Experience counts: integrating spearfishers' skills and knowledge in the evaluation of biological and ecological impacts. Fish Manag Ecol 24(2): 95–102. https://doi.org/10.1111/fme.12206

Fischer J, Jorgensen J, Josupeit H, Kalikoski D, Lucas CM (2015) Fishers' knowledge and the ecosystem approach to fisheries: applications, experiences and lessons in Latin America. In: FAO fisheries and aquaculture technical paper No. 591. FAO, Rome, p 278

Fitzgerald CJ, Delanty K, Shephard S (2018) Inland fish stock assessment: applying datapoor methods from marine systems. Fish Manag Ecol 25:240–252. https://doi.org/10.1111/fme.12284

McCune B, Grace JB (2002) Analysis of ecological communities. MjM Software Design, Oregon

Mukherjee N, Huge J, Sutherland WJ, McNeill J, Opstal MV, Dahdouh Guebas F, Koedam N (2015) The Delphi technique in ecology and biological conservation: applications and guidelines. Methods Ecol Evol 6(9):1097–1109. https://doi.org/10.1111/2041-210X.12387

van Densen WLT (1990) Proc. Symp. European Inland Fisheries Advisory Commission, Göteborg, Sweden, p. 301–312

Concept of Sampling Methodologies and Their Applications

2

Key Topics
- The concept of sampling
- Water and fish sampling
- Different sampling methods
- Ecological sampling methods

As complete enumeration of a population is very often untenable because of time, cost and various other constraints, a part of the population is studied to draw inferences on the population; and this process is known as sampling. According to Mildred Parton, 'Sampling method is the process or know-how of drawing a definite number of the individuals, cases or the observations from a particular universe, selecting part of a total group for investigation'. The advantages of sampling are very accurate, reliable, consuming less time and economical. Sampling has also some disadvantages, such as inadequacy of the samples, chances for bias, problems of accuracy, difficulty in representative sample, chances of sampling errors, etc. Population is defined as the collection of all individuals or items of a particular characteristic of interest in a certain geographical area, and sample is a subset of the population. In some research trials, it may be possible to take the measurement of all the objects, but most of the time, monitoring the entire population becomes too costly and time-consuming. For example, catching and measuring length and weight of 40 fishes or 80 fishes individually from a few experimental tanks is possible, but catching 500 fishes from a big pond or reservoir and measuring their length and weight is difficult. When it is supposed to find the average weight of fish in a big pond or reservoir, which is impossible, then manageable portions of the population are taken as a sample. Sample numbers vary widely and depend upon the population situation, e.g. 1%, 5%, 10%, 25% or even more. The larger sample size provides the higher accuracy and reliability, but simultaneously it also increases the cost. So,

B. K. Das et al., *Concept Building in Fisheries Data Analysis*,
https://doi.org/10.1007/978-981-19-4411-6_2

sample size should be minimal but taken randomly so that all individuals of the population have an equal chance of being selected and represent the population. The distinct proper sampling techniques are followed to make the samples representative of the population in different situations. There are two types of sampling: non-probability sampling and probability sampling. Non-probability sampling is a sampling technique where the samples are gathered in a process that does not give all the individuals in the population equal chances of being selected, e.g. convenience sampling, purposive or judgemental sampling, etc. Probability sampling is a sampling technique where the samples are gathered in a process that gives all the individuals in the population equal chances of being selected, e.g. random sampling, stratified sampling, systematic sampling and cluster sampling.

2.1 Some Definitions

Some basic definition and concepts used in sampling procedure are as follows:

- Sampling unit: Elementary units or group of such units which besides being clearly defined, identifiable and observable are convenient for purpose of sampling are called sampling units. In a fish catch and resource survey, a pond (water body) or a group of ponds owned or operated by a household may be considered as the sampling unit. In an industrial survey of estimating the total production of an item, an industry manufacturing that particular item is a sampling unit.
- Sampling frame: A list of all the sampling units belonging to the population to be studied with their unique identification particulars or a map showing the boundaries of the sampling units is known as sampling frame. Examples of a frame are a list of water resources and a list of suitable area segments like villages in India. The information on wetlands (e.g. area, location, etc.) of states. The frame should be up to date and free from errors of omission and duplication of sampling units.
- Random sample: One or more sampling units selected from a population according to some specified procedures are said to constitute a sample. The sample will be considered as random or probability sample, if its selection is governed by ascertainable laws of chance. In other words, a random or probability sample is a sample drawn in such a manner that each unit in the population has a predetermined probability of selection. For example, if a population consists of the N sampling units (clearly defined, identifiable and observable) U_1, U_2,. . ., U_i,. . ., U_N, then we may select a sample of n units by selecting them unit by unit with equal probability for every unit at each draw with or without replacing the sampling units selected in the previous draws.
- Population parameters: Suppose a finite population consists of the N units $U_{1,2}$ and let Y_i be the value of the variable Y, the characteristic under study, for the ith unit ($i = 1, 2,. . ., N$). For instance, the unit may be a wetland, and the characteristic under study may be the fish production. Any function of the values of all the population units (or of all the observations constituting a population) is known as

a population parameter or simply a parameter. Some of the important parameters usually required to be estimated in surveys are the population total, $Y = \frac{1}{n}\sum_{i=1}^{n} Y_i$, and the population mean.

- Statistic, estimator and estimate: Suppose a sample of N units is selected from a population of N units according to some probability scheme, and let the sample observations be denoted by $y_1, y_2, \ldots, y_n$. Any function of these values which is free from unknown population parameters is called a statistic. An estimator is a statistic obtained by a specified procedure for estimating a population parameter. The estimator is a random variable, and its value differs from sample to sample, and the samples are selected with specified probabilities. The particular value, which the estimator takes for a given sample, is known as an estimate.
- Sample design: A clear specification of all possible samples of a given type with their corresponding probabilities is said to constitute a sample design. For example, suppose we select a sample of n units with equal probability with replacement and the sample design consists of N^n possible samples (taking into account the orders of selection and repetitions of units in the sample) with $\frac{1}{N^n}$ as the probability of selection for each of them, since in each of the n draws any one of the N units may get selected. Similarly, in sampling N units with equal probability without replacement, the number of possible samples (ignoring orders of selection of units) N_{C_n} is and the probability of selecting each of the samples is $\frac{1}{N_{C_n}}$.
- Unbiased estimator: Let the probability of getting the ith sample be P_i, and let t_i be the estimate, that is, the value of an estimator t of the population parameter θ based on this sample ($i = 1, 2, \ldots, M_o$), M_0 being the total number of possible samples for the specified probability scheme. The expected value or the average of the estimator t is given by

$$E(t) = \sum_{i=1}^{M_0} iP_i$$

An estimator t is said to be an unbiased estimator of the population parameter θ; its expected value is equal to θ irrespective of the y values. In the case the expected value of the estimator is not equal to the population parameter, the estimator t is said to be a biased estimator of θ. The estimator t is said to be positively or negatively biased for the population parameter according as the value of the bias is positive or negative.

2.2 Simple Random Sampling

The simple random sampling is the most commonly used sampling method; it selects samples randomly from a well-defined population. More technically, a unit/individual is chosen entirely by chance, such that each unit/individual has the same probability of being selected at any stage during the sampling process. There are two simple random sampling methods: simple random sampling with replacement (SRSWR) and simple random sampling without replacement (SRSWOR). The SRSWR allows a population unit to be selected more than once, but SRSWOR selects any population unit only once. When a small sample is selected from a large population size, then SRSWOR is approximately the same as the SRSWR. However, the necessity of SRSWR arises seldom in the case of inland fisheries assessment; thereby, this chapter describes only the SRSWOR. Hence, simple random sampling always refers to the SRSWOR throughout this chapter. The estimates of the population mean and its variance are given as below.

Let us assume that a sample of n individuals is selected by simple random sampling with the values $x_1, x_2, \ldots, x_n$ of a population of size N. The unbiased estimate of the population mean is given by

$$\overline{x} = \frac{1}{n}\sum\nolimits_{i=1}^{n} x_i \tag{2.1}$$

The variance of the sample mean is estimated by the formula:

$$\text{var}(\overline{x}) = \left(\frac{1}{n} - \frac{1}{N}\right)S^2 \tag{2.2}$$

where $S^2 = \frac{1}{n-1}\sum_{i=1}^{n}(x_i - \overline{x})^2$ is the sample variance, which is the unbiased estimator of the population $S^2 = \frac{1}{n-1}\sum_{i=1}^{n}\left(X_i - \overline{X}\right)^2$. The standard error (SE) of the sample mean is given by

$$SE(\overline{x}) = \sqrt{\text{var}(\overline{x})} \tag{2.3}$$

when N is large compared to n, then the variance and the SE of the mean become

$$\text{var}(\overline{x}) = \frac{s^2}{n} \tag{2.4}$$

$$SE(\overline{x}) = \frac{s}{\sqrt{n}} \tag{2.5}$$

So, sample variance of $\overline{x}$ depends upon the sample size (=n) and variability present in the population. All the estimates for the population total (X) are

$$\widehat{X} = N\overline{x};\ Var\left(\widehat{X}\right) = N^2 Var(\overline{x});\ \text{and}\ S.E.\left(\widehat{X}\right) = \sqrt{\text{var}\left(\widehat{X}\right)}.$$

Example 2.1 illustrates the step-by-step computation of all the formulas mentioned above.

Example 2.1 In a particular water body, fishing happens daily. Ten days were selected in a month (of 30 days) by simple random sampling without replacement

Table 2.1 Fish landings of 10 randomly selected days

Day	1	2	3	4	5	6	7	8	9	10
Catch (kg)	732	1021	631	428	376	673	512	437	801	636

(SRSWOR), and the fish catches (in kg) on these sampled days were recorded in Table 2.1. Estimate the average fish catch per day and the total fish catch of the month along with their standard errors.

Solution The catch per day is estimated by using Formula 2.1 for sample mean as

$$\begin{aligned}\overline{x} &= \frac{1}{n}\sum_{i=1}^{n} x_i \\ &= \frac{1}{10} \times 6247 \\ &= 624.7\end{aligned}$$

and the sample variance is

$$\begin{aligned}S^2 &= \frac{1}{n-1}\sum_{i=1}^{n}(x_i - \overline{x})^2 \\ &= \frac{1}{(10-1)} \times 350,624.1 \\ &= 38,958.23\end{aligned}$$

So, the estimate of the variance is computed from Formula 2.2 as

$$\begin{aligned}\mathrm{var}(\overline{x}) &= \left(\frac{1}{n} - \frac{1}{N}\right)S^2 \\ &= \left(\frac{1}{10} - \frac{1}{30}\right) \times 38,958.23 \\ &= 2597.216\end{aligned}$$

The standard error of the mean is $SE(\overline{x}) = \sqrt{\mathrm{var}(\overline{x})} = \sqrt{2597.216} = 50.963$. The estimate of total catch during the month and its standard error is computed as

$$\begin{aligned}&\widehat{X} = N\overline{x} = 30 \times 624.7 = 18741 \\ &SE\left(\widehat{X}\right) = N \times SE(\overline{x}) = 30 \times 50.963 = 1528.887\end{aligned}$$

2.3 Systematic Sampling

The method of sampling in which only the first unit is selected at random and the rest is being selected according to a predetermined pattern is known as 'systematic sampling'. Let us assume that in a population of size N, the units are serially numbered as 1 to N, and a sample of size n is desired to be drawn from it. Now assume that $N = nk$ where k is an integral part of N; now $k = N/n$ is called the sampling interval, and the sample drawn with this interval is called a 1 in k sample. Drawing a systematic sample consists of selecting a unit at random from the first k units and then selecting every kth unit in the population thereafter. At a particular landing centre, for example, data on fish landings were recorded during a month (of 30 days) on 10 days selected by systematic sample with a sampling interval of 3 days. The landings (in tons) of the 10 sampling days are given below:

For example, if there is a population of size N= 30 days and it is required to draw a sample of size $n = 10$, in this case $k = 30/10 = 3$. Hence, this is one in three samples. Therefore, draw a unit at random from the first three units. Suppose this turns out to be the second unit, and then the sample consists of the second unit and every fifth unit thereafter, i.e. units with serial number 2, 5, 8,. . ., and 29 days constitute a sample of ten sample size. When $N = nk$, the sample mean is unbiased, and when $N \neq nk$, the sample mean gives biased estimate. When $n > 50$ the bias is negligible and can be ignored. An alternative method of drawing a sample when $N \neq nk$, which gives unbiased estimate of the population mean, is also available. For details readers may refer to Cochran (1977).

2.3.1 Advantages and Disadvantages Advantage

- Easy to draw and requires less time
- Samples are spread out more evenly over the population
- It is likely to be more precise than the simple random sampling for many populations.

Disadvantage

- There is no reliable method for estimating the variance of a systematic sample from the sample data.
- Systematic sampling has to be used carefully when there is a periodicity in the population, as its efficiency depends upon the choice of the sampling interval.

Example 2.2 In a particular landing centre, during a month (consisting of 30 days), a systematic sample of 10 days is selected with a sampling interval of 3 days. The fish landings (in kg) on those selected days were recorded in Table 2.2.

1. Estimate the average fish landing per day at the landing centre along with its standard error.

Table 2.2 Fish landings for 10 days

Day	1	2	3	4	5	6	7	8	9	10
Landings (kg)	23	45	32	67	86	72	55	76	87	29

2. Give an estimate of the total fish landings of the centre during the month and its appropriate standard error.

Solution

1. The average fish landings per day is given by

$$\bar{x}_{sys} = \frac{1}{n}\sum_{i=1}^{n} x_i$$
$$= \frac{1}{10} \times 572 = 57.2 \text{ kg}$$

The variance of the mean (of systematic sampling) is estimated by

$$\widehat{Var}(\bar{x}_{sys}) = \frac{(k-1)}{2nk(n-1)}\sum\nolimits_{i=1}^{n-1}\left(x_{(i+1)} - x_{(i)}\right)^2 \tag{2.6}$$

$$= \frac{(3-1)}{2 \times 10 \times 3 \times (10-1)} \times 6650 = 24.63$$

The standard error (SE) is computed as

$$SE(\bar{x}) = \sqrt{(Var(\bar{x}))}$$
$$= \sqrt{24.63} = 4.963$$

2. The estimate of the total fish landings for the centre for the month is given by

$$\widehat{X}_{sys} = 30 \times \bar{x}_{sys}$$
$$= 30 \times 57.2$$
$$= 1716$$

The standard error of the estimate of the total fish landings is given by

$$SE\left(\widehat{X}_{sys}\right) = N \times SE(\bar{x})$$
$$= 30 \times 4.963$$
$$= 148.89$$

2.4 Stratified Random Sampling

Note that the variance of sample mean depends on the size of the sample and the variability of the population. Therefore, the only way of increasing the precision of an estimate apart from the size of the sample is to devise sampling procedures which will effectively reduce the variability. One such procedure known as 'stratified sampling' consists of dividing the population into 'classes' or 'strata', each relatively homogeneous and drawing random samples of known sizes, one each from different strata. Then the estimates are made for each of the strata and combined by a proper weight to obtain the estimate for the whole population. The variance of this estimate is obtained by combining the variances of the estimates within each stratum. This combined estimate of variance will be small as within stratum variances will tend to be small as each of them comes from relatively homogeneous stratum.

Stratified random sampling method assumes that the structure of the population necessary to demarcate the strata is known, which may not be true in many situations. For better planning and execution of the survey work, stratification is sometimes done based on geographical proximity.

2.4.1 Estimation of Mean and Variance

If the population under study is divided into k strata with sizes N_1, N_2,..., N_k, respectively, then estimate of population mean is given by

$$\overline{x} = \sum_{i=1}^{k} \frac{N_i x_i}{N}$$

where $\overline{x}_1, \overline{x}_2, \ldots, \overline{x}_k$ are the means of 1st, 2nd,..., kth stratum, respectively, $N = \sum_{i=1}^{k} N_i; \overline{x}_i = \frac{1}{N_i}\sum_{j=1}^{k} x_{ij}$, where n is the sample size of the ith stratum. The variance of the sample mean is given by

$$Var(\overline{x}) = \frac{1}{N^2}\sum_{i=1}^{k} N_i^2 Var(\overline{x}_i) \qquad (2.7)$$

where $Var(\overline{x}_i) = \left(\frac{1}{n_i} - \frac{1}{N_i}\right)S_i^2; S^2 = \frac{1}{(N_i - 1)}\sum_{i=1}^{k}(x_i - \overline{x})^2$. The estimate of population total (X) and its variance are given by $\widehat{X} = N\overline{x}$ and $Var\,\widehat{X} = N^2 Var(\overline{x})$, respectively.

2.4.2 Allocation of Sample Size

The advantage of stratified sampling can be increased by allocating the sample size n into various strata in the best proportion. The following are the important methods of allocation of samples to different strata.

1. Equal samples for each stratum: The total sample size n is divided equally among all the strata, i.e. for the *i*th stratum, if there are $n_i = \frac{n}{k}$ strata.
2. Proportional allocation: In this method the total sample size n is allocated to different strata in proportion to their size, i.e. for the *i*th stratum.
3. Optimum allocation: In some cases it may be required to conduct a sample survey with a fixed budget, but with varying costs of selecting the sampling units from different strata. In such situations the allocation of the number of units to each stratum is done with a view to minimizing the sampling variance for the given fixed cost, or the cost of survey is minimum for the specified value of the sampling variance. Allocation of the sample for the *i*th stratum is given by

$$n_i = n\left(\frac{\frac{N_iS_i}{\sqrt{C_i}}}{\sum_{i=1}^{k}\frac{N_iS_i}{\sqrt{C_i}}}\right)$$

where N_i is the size of the *i*th stratum, S_i is the population standard deviation for the N_i stratum and C_i is the cost of obtaining a single observation from the *i*th stratum.

4. Neyman allocation: This method of allocation is used when the cost of selecting the sampling unit does not differ from stratum to stratum. The sample size for the *i*th stratum is given by

$$n_i = n\left(\frac{N_iS_i}{\sum_{i=1}^{k}N_iS_i}\right)$$

This indicates that the total sample size n is allocated in proportion to N_iS_i, that is, take more individuals from the strata that are large and are highly variable. Sometimes it may so happen that the sample size n_i estimated by Neyman allocation may be larger than the corresponding N_i of the *i*th stratum. Such a type of situation arises only when overall sampling fraction $\left(f = \frac{n}{N}\right)$ is substantial and one stratum is much variable than the others. The procedure that can be adopted in such a situation is outlined in Cochran (1977).

2.4.3 Advantages of Stratified Random Sampling

There are many advantages with stratified random sampling, and the important ones are given below:

1. Stratified random sampling gives better cross section of the population than simple random sampling.
2. Precision of the estimated character is likely to be higher in stratified random sampling than in simple random sampling.

3. For physical or administrative reasons, it is easier to collect the data using this sampling technique.
4. This procedure will be especially useful when sampling problems differ in different parts of the population.

Example 2.3 In a fishing village consisting of 200 families, fishers are engaged in fishing activities for their additional income. These families are broadly classified as low (100 nos), medium (60 nos) and high (40 nos) income earning families based on their additional income. A random sample of 10, 6 and 4 families from low-, medium- and high-income categories was selected, respectively, and the additional income (in Rs.) for the selected families was recorded as follows:

Low: 155, 162, 175, 200, 197, 187, 176, 188, 176, 173
Medium: 245, 233, 256, 261, 272, 277
High: 356, 340, 375, 339
Estimate the additional income (in Rs.) per family along with its standard error.

Solution Example 2.3 is a case of stratified sampling with three strata designated to low-, medium- and high-income category. The measured population characteristic is the additional income of family, the sampling unit.

For the low-income category: $N_1 = 100$ and $n_1 = 10$.

$$\begin{aligned}\overline{x} &= \frac{1}{n_i}\sum_{j=1}^{n_1} x_{1j}\\ &= \frac{1789}{10}\\ &= 178.9\end{aligned}$$

$$\begin{aligned}S_1^2 &= \frac{1}{n_i - 1}\left[\sum_{j=1}^{n_1} x_{1j}^2 - \frac{\left(\sum_{j=1}^{n_1} x_{1j}\right)^2}{n_1}\right]\\ &= \frac{1}{(10-1)}\left[321,897 - \frac{3,200,521}{10}\right]\\ &= 204.989\end{aligned}$$

$$\begin{aligned}Var(\overline{x}_1) &= \left(\frac{1}{n_1} - \frac{1}{N_1}\right)s_1^2\\ &= \left(\frac{1}{10} - \frac{1}{100}\right) \times 204.989\\ &= 18.449\end{aligned}$$

For the medium-income group: $N_2 = 60$; $n_2 = 6$

$$\overline{x}_2 = \frac{1544}{6} 257.333$$

$$s_2^2 = \frac{1}{6-1}\left[398,684 - \frac{(1544)^2}{6}\right]$$

$$= 272.267$$

$$Var(\overline{x}_2) = \left(\frac{1}{6} - \frac{1}{60}\right) \times 272.267$$

$$= 40.84$$

For the high-income group: $N_3 = 40$; $n_3 = 4$

$$\overline{x}_3 = \frac{1410}{4} = 352.5$$

$$s_3^2 = \frac{1}{4-1}\left[497,882 - \frac{(1410)^2}{4}\right]$$

$$= 285.667$$

Hence, the mean additional income per family and its standard error are given by

$$\overline{x} = \frac{1}{N}\sum_{i=1}^{3} N_i\overline{x}_i$$

$$= \frac{47,430}{200} = 237.15$$

$$Var(\overline{x}) = \frac{1}{N^2}\sum_{i=1}^{3} N_i^2 \times Var(\overline{x}_i)$$

$$= \frac{N_1^2 \times Var(\overline{x}_1) + N_2^2 \times Var(\overline{x}_2) + N_3^2 \times Var(\overline{x}_3)}{N^2}$$

$$= \frac{(100)^2 \times 18.449 + (60)^2 \times 40.84 + (40)^2 \times 64.275}{(200)^2}$$

$$= 10.859$$

$$SE(\overline{x}) = \sqrt{Var(\overline{x})} = \sqrt{10.859} = 3.295$$

2.5 Cluster Sampling

If the basic sampling unit in a population is to be found in groups of clusters, then from the operational point of view, the sampling may be carried out by selecting a sample of clusters and observing all the units of each selected cluster. This type of sampling is known as cluster sampling.

It is less costly than simple random sampling. However, cluster sampling is less efficient than simple random sampling due to the fact that individual units within a cluster tend to be similar. Efficiency of cluster sampling can be increased by increasing the size of the sample. In many situations, it costs less to take considerably larger cluster samples than to take smaller simple random samples with the same precision. Hence, it is generally expected that efficiency per unit cost will be more in cluster sampling than in simple random sampling.

Choice of cluster sampling in fish catch surveys is of immense use due to the following two major advantages:

1. Sometimes there is no sampling frame available to base a sample selection. Such a list is usually expensive to obtain, whereas it is relatively easy to obtain a list of cluster of units.
2. Cluster sampling may also be desirable if the population is such that sampling costs may be reduced by selecting adjacent units. For example, it may be cheaper to select some villages and sample good proportion of ponds and tanks or fishing units rather than to take simple random sample of ponds or fishing units in the study area.

2.5.1 Estimation of Population Mean

Let N and n denote, respectively, the number of clusters in the population and in the sample, respectively. Let M_i denote number of units in the ith cluster. For simplicity consider the case of equal clusters, i.e. $M_i = M$ for all i. Then an unbiased estimate of the population mean is given by

$$\overline{x} = \frac{1}{n}\sum \overline{x}_i$$

where $\overline{x}_i$ is the mean of the ith cluster. The estimate of the variance is given by

$$Var(\overline{x}) = \left(\frac{1}{n} - \frac{1}{N}\right)\frac{1}{N-1}\sum_{i-1}^{n}(\overline{x}_i - \overline{x})^2$$

The estimate of the population total and its variance is

$$X = N(\overline{x}_i)$$
$$Var(X) = N^2 Var(\overline{x})$$

Example 2.4 District catch estimation A particular district has got 27 clusters of fishing villages—each cluster having 5 villages. A random sample of 9 clusters were selected from these 27 clusters, and monthly fish catch (in kg) from all the villages from the selected clusters was recorded as shown in Table 2.3.

1. Estimate the average fish catch (in kg) per village during the sampled month along with its standard error.

Table 2.3 Monthly fish catch (in kg) for sampled clusters

Cluster	Village 1	Village 2	Village 3	Village 4	Village 5	Mean
1	43	65	78	87	78	70.2
2	45	46	65	66	47	53.8
3	12	34	32	89	71	47.6
4	56	10	12	52	38	33.6
5	65	76	89	31	42	60.6
6	72	32	55	36	39	46.8
7	14	55	43	48	50	42.0
8	27	67	27	48	52	44.2
9	33	89	33	65	76	59.2

2. Estimate the total fish catch in the district during the month along with its standard error.

Solution

1. The estimate of the average fish catch village and its standard error is:

$$\overline{x} = \frac{1}{n}\sum \overline{x}_i$$
$$= \frac{458}{9}$$
$$= 50.89$$

$$Var(\overline{x}) = \left(\frac{1}{n} - \frac{1}{N}\right)\frac{1}{N-1}\sum_{i=1}^{n}(\overline{x}_i - \overline{x})^2$$

$$Var(\overline{x}) = \left(\frac{1}{9} - \frac{1}{27}\right) \times \frac{1}{9-1} \times 994.946$$
$$= 9.213$$

$$\text{Standard Error } (SE) = \sqrt{Var(\overline{x})}$$
$$= \sqrt{9.213}$$
$$= 3.035$$

2. Estimate of the total catch (X) and its standard error in the district during the month are as follows:

$$= (27 \times 5) \times 50.89 = 6870 \text{ kg}$$
$$(\overline{X}) = \text{Total number of villages in the district} \times SE(\overline{x})$$
$$= (27 \times 5) \times 3.035$$
$$= 409.757$$

2.6 Multistage Sampling

It is logical to expect that the efficiency of the estimator can be increased by distributing elements over a larger number of clusters and surveying only a sample of units in each selected cluster instead of completely enumerating all the units in the selected clusters. This logic gives rise to a new sampling procedure.

The procedure of sampling, which consists in first selecting the clusters and then randomly choosing a specified number of units from each selected cluster, is known as two-stage sampling. This is also called sub-sampling. In such sampling designs, clusters which form the units of sampling at the first stage are called first-stage units (fsu's) or primary stage units (psu's), and the elements within the clusters are called second-stage units (ssu's). This procedure can be generalized to three or more stages and is termed as multistage sampling. For example, in catch estimation surveys for estimating yield of fish in a district, a block may be considered as a first-stage unit, villages within blocks as the second-stage units and the pond within village as the ultimate unit of sampling.

Multistage sampling has been found to be very useful in practice, and this procedure is being commonly used in large-scale surveys. This sampling procedure is a compromise between cluster sampling and direct sampling of units. Further, this design is more flexible as it permits the use of different selection procedures at different stages. It may also be mentioned that multistage sampling may be the only choice in a number of practical situations where a satisfactory sampling frame of ultimate-stage units is not readily available and the cost of obtaining such a frame is large and time-consuming.

Figure 2.1 depicts graphical presentation of a multistage (with three-stage) sampling scheme. For the selection of sample under multistage sampling when the population is too large, sampling should be done a few times instead of just once. First-stage sampling is random. At the second stage, samples can be taken from two opposite sides (up and down) out of four quarters of a circle. Then at the third stage, again, selecting opposite sides (left right) could produce a good sample.

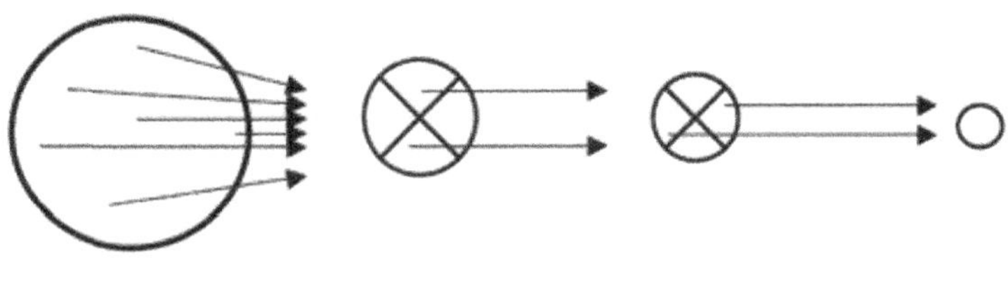

Fig. 2.1 Schematic diagram of a three-stage sampling

2.7 Sample Size

In most cases, sample size may mean the number of animals or objects that represent the whole population of any experimental unit. Most researchers often decide based on the percentage, e.g. 20%, 10%, 5% or 3%. But when there are situations in which researchers need to make decisions about sample size from sufficiently large populations, e.g. tens of thousands, hundreds of thousands or even millions, then a percentage formula would not work because even 0.1% would result in a large number that is not manageable. Two methods have been found to determine sample size.

2.8 Margin of Error

The margin of error is a statistic expressing the amount of random sampling error in a survey results. In statistics, the two most important ideas regarding sample size and margin of error are, first, sample size and margin of error have an inverse relationship and, second, after a point, increasing the sample size beyond what you already have gives you a diminished return because the increased accuracy will be negligible. The relationship between margin of error and sample size is simple: as the sample size increases, the margin of error decreases (see Fig. 2.2). This relationship is called an inverse because the two move in opposite directions. If you think about it, it makes sense that the more information you have, the more accurate your results are going to be (in other words, the smaller your margin of error will get). (That assumes, of course, that the data were collected and handled properly).

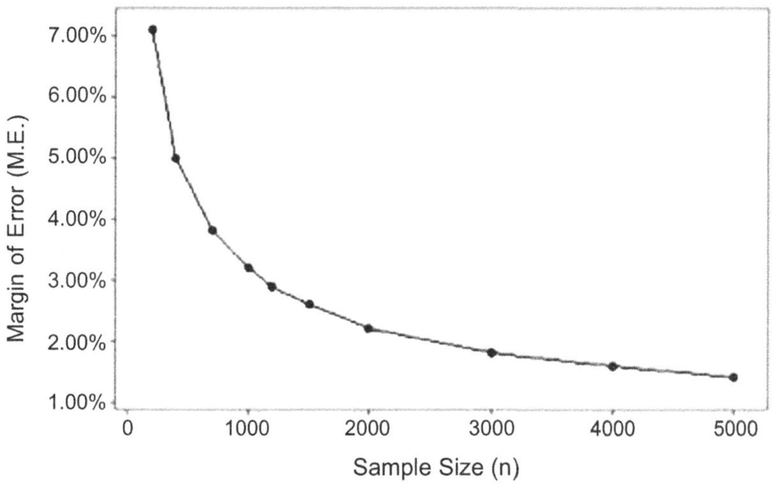

Fig. 2.2 Margin error and sample size relationship

Table 2.4 z^* values at different confidence levels

Percentage confidence	z^* value
80	1.28
90	1.645
95	1.96
98	2.33
99	2.58

Example 2.5 Suppose that to know the performance of a selective new fish species growth in a given region we sampled a 1000 fishermen of that region in a state, and the results show that 520 fishermen (52%) think that the new fish species growth is good, compared to 48% who don't think so. First, assume you want a 95% level of confidence, so you find z^* using Table 2.4.

From Table 2.4, you find that $z^* = 1.96$. The number of fishermen in the sample who said good growth of the fish was found to be 520. This means that the sample proportion is $\frac{520}{1000} = 0.52$. (The sample size, n, was 1000). The margin of error for this question is calculated in the following way:

$$\begin{aligned} z^*\sqrt{\frac{p(1-p)}{n}} &= 1.96 \times \sqrt{\frac{0.52 \times 0.48}{1000}} \\ &= 1.96 \times 0.0158 \\ &= 0.0319 \end{aligned}$$

According to this data, you conclude with 95% confidence that 52% of all fishermen approve the new fish species, plus or minus 3.1%.

Using the same formula, you can look at how the margin of error changes dramatically for samples of different sizes. Suppose in the fishermen approval poll that n was 500 instead of 1000. Now the margin of error for 95% confidence is $1.96 \times \sqrt{\frac{0.52\times0.48}{500}} = 0.0438$ which is equivalent to 4.38%. If n is increased to 1500, the margin of error (with the same level of confidence) becomes $1.96 \times \sqrt{\frac{0.52\times0.48}{1500}} = 0.0253$ or 2.53 %. Finally, when $n = 2000$, the margin of error is $1.96 \times \sqrt{\frac{0.52\times0.48}{2000}} = 0.0219$ or 2.19%.

Looking at these different results, you can see that larger sample sizes decrease the margin of error, but after a certain point, you have a diminished return. Each time you survey one more person, the cost of your survey increases, and going from a sample size of, say, 1500 to a sample size of 2000 decreases your margin of error by only 0.34% (one third of one percent!)—from 0.0253 to 0.0219. The extra cost and trouble to get that small decrease in the margin of error may not be worthwhile. Bigger isn't always that much better!

2.9 Fish Sampling

A researcher can weigh the whole group (batch weight per replicate) and divide by that number to estimate the mean weight or sample a certain number of fish to weigh individually and compute the sample average as a representative weight of the whole replicate group. The first method gives a more accurate mean; however, in many cases, it is not possible to weigh all of the fish. In such cases, sampling of a certain number of fish is necessary. Live fish sampling is quite difficult to do, and it is difficult to determine whether a proper random sample can be drawn because fish move fast from one corner to another. If the experimental units, e.g. ponds or tanks, are quite big, although it is difficult, it should be kept in mind that samples need to be taken from the middle parts as well as from every corner. Most researchers do carry out the intermediate samplings but do not use data of intermediate samplings and use only final or end point data for statistical analysis. If there are intermediate samplings, the data collected can be valuable and informative to see the trend over the period rather than just the result at the end. In such a case, time should be considered as block while performing statistical analysis. Therefore, separating the block effect would increase the reliability of the test or decrease the chance of committing an error.

2.10 Water Sampling and Monitoring

The environmental factors to be considered are water temperature, dissolved oxygen (DO), pH, turbidity, chlorophyll-*a*, ammonia, nitrate, nitrite concentrations, etc. These parameters cyclically fluctuate daily in a pond. Monitoring of water quality, early in the morning between 06:00 and 06:30 h for the lowest and in the afternoon between 14:00 and 16:00 h to record the highest levels, is important, especially for temperature and DO over the experimental period. Most water quality parameters usually vary with depth; therefore, sampling or measurement should also be taken at the depths of 20 and 50 cm.

2.11 Ecological Sampling Methods

An ecologist is always interested to know about the density (number of individual per unit area) and dispersal (the pattern of distribution in space) of fish species, planktonic species and benthic species in the ecosystem. So it would be useful if there are some ways that we could get an accurate representation of some spatial characteristics of the population without having to map every organism. This can be achieved by selecting some samples of species from the population.

2.11.1 Some Spatial Sampling Methods

There are three general types of sampling methods used to select individuals from a spatially dispersed population, quadrat method, transect method and plotless techniques, as discussed below.

Quadrat Method A quadrat is a frame (usually a square or a circle) of the known area used to isolate a subset of the population. This subset will comprise one sample. Quadrats come in various sizes (and shapes)—the features of the sampling organism determine quadrat size. Quadrats typically consist of a square frame; the most frequently used size is 15 m^2, e.g. a square quadrat of each side one meter is used for benthic population sampling. The purpose of using a quadrat is to enable comparable samples of consistent size and shape. Rectangular quadrats and even circular quadrats have been used in some surveys. It may, however, be better to stick to the traditional square frame unless there are very good reasons not to because this yields data that is more readily comparable to other published research. For instance, you cannot compare data obtained using a circular quadrat, with data obtained using a square quadrat. The difference in the shape of the sampling units will introduce variations in the results obtained.

Quadrat sampling is straightforward to implement. Typically, a quadrat is placed in the sampling area—the habitat of the species of interest and all the individuals within the quadrat are measured. Precisely, what the sampling area is, how its limits are determined and how the quadrats are placed are all very important points in using this method. Quadrats are most useful when the area is fairly uniform and movement within the area is easy.

Quadrats are most often used for sampling but are not the only type of sampling units. It depends on what you are sampling. If you are sampling aquatic microorganisms or studying water chemistry, then you will most likely collect water samples in standardized bottles or containers. If you are looking at parasites on fish, then an individual fish will most likely be your sampling unit. In these cases, the sampling units will not be of standard size.

This problem can be overcome by using a weighted mean, which takes into account different sizes of sampling unit, to arrive at the mean number of organisms per sampling unit.

Transect Method The transect method is most useful when the area to be sampled is zoned in some way or has some sort of gradients running through it. There are two methods of transect sampling.

1. Line transect method: A line transect is a path along which one counts or records the occurrence of the species under study. It is useful when the area to be sampled is zoned in some way or has some sort of gradients running through it. The transect line is randomly laid along the margins of the zones (see Fig. 2.3). The position of the transect line is very important, and it depends on the direction of

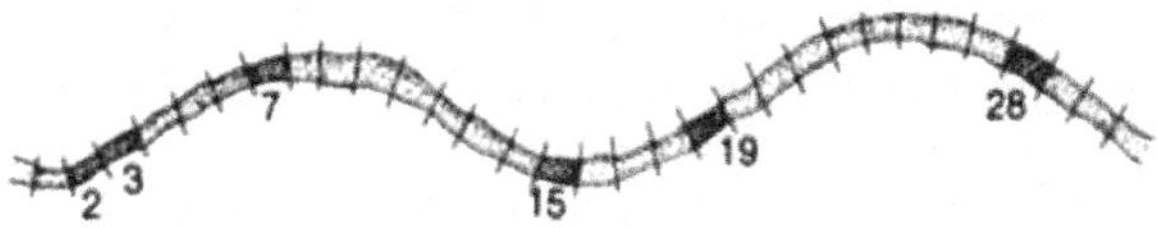

Fig. 2.3 Selection of sampling site using line transect method

the environmental gradient you wish to study. Transects can either be continuous with the whole length of the line being sampled, or samples can be taken at particular points along the line, for example, every metre or every other metre.

A line transect is carried out by unrolling the transect line along the gradient identified. The species touching the line may be recorded along the whole length of the line (continuous sampling). Alternatively, the presence or absence of species at each marked point is recorded (systematic sampling). If the slope along the transect line is measured as well, the results can then be inserted onto this profile.

Once the line has been randomly laid, points along the line are selected (or quadrats could be used too). The points might be randomly or uniformly placed, depending on the question being asked. In this way the distribution of a species across (or within) the environmental gradient can be determined.

Note: *Line transect is used in acoustic survey; water and soil sites are selected in the river.*

2. Belt/strip transect method: It may be considered as a widening of the line transect to form a continuous belt or series of quadrats. It gives information on abundance as well as presence or absence of species. In this method, the transect line is laid out across the area to be surveyed, and a quadrat is placed on the first marked point on the line. The species inside the quadrat are then identified and their abundance estimated. Quadrats are sampled along the transect line, at each marked point on the line or at some other predetermined interval (or even randomly) if time is short.

   Note: *Experimental fishing into the river can be done using belt transect method to know the abundance of species.*

   The main utilities of line and belt transect are as follows:

   Utilities of line transect sampling

- Where time is limited, a line transect method can be carried out quickly.
- Useful for visual illustration of species change along line.
- To show species range along the line. This will generally show only species occurrences but not abundance of it.
- Very useful for measuring water velocity and depth in a river sampling

   Utilities of belt transect sampling

- A belt transect will supply more data than line transect.
- It gives data on the abundance of individual species at different points along the line, as well as on their range.
- To show how the abundance of each individual species changes within its range.
- Belt transect data will allow the relative dominance of species along the line to be determined.

Plotless Techniques It is particularly useful in areas where carrying large quadrats would be difficult to manage, e.g. to find a diseased fish. It is also most suited to habitats that are relatively uniform. Plotless methods do not use a defined area (like a quadrat), nor are they arranged linearly as in transect methods; rather, they use distances between a point and an individual or between two individuals.

In this technique each sample consists of a pair of measurements, d_1 is the distance from the randomly selected point to the closest individual of interest, and d_2 is the distance to the nearest neighbour. To obtain the next sample, a new random point is chosen, and another pair of measurements is made. Unlike quadrat methods, this method is not that useful for determining density. It is however useful for determining the dispersion pattern of the species/organisms.

2.11.2 Density and Dispersion Patterns

As noted earlier, the number of individuals per unit area is termed the density, and the units are things, such as fish km^{-1}, zooplankton m^{-3}, parasites fish^{-1}, etc. Dispersion is the pattern of the distribution of organisms in space. There are three basic dispersion patterns: random, regular and contagious.

A random dispersion pattern means that there is an equal probability of an individual occurring at any point in the habitat and that the presence of an individual does not influence the probability of occurrence of another individual.

Regular dispersion, indicated by more even spacing that would be predicted by a random dispersion, may suggest territoriality or some limiting resource.

Contagious dispersion patterns are those where the presence of an individual increases the probability of finding another one nearby.

2.11.3 Determining the Dispersion Pattern

For quadrat data, the mean $(= \bar{x})$ and variance $(=s^2)$ are calculated taking the total number of individuals counted in each sample as the individual samples. The detection of dispersion patterns is based on the following series of possible value of ratio of variance to mean, as follows:

$\frac{s^2}{\bar{x}} = 1, \quad \Rightarrow$ random pattern of species distribution

$\frac{s^2}{\bar{x}} < 1, \quad \Rightarrow$ regular pattern of species distribution

$\frac{s^2}{\bar{x}} > 1, \quad \Rightarrow$ contagious pattern of species distribution

To test whether the difference is large enough to be considered significant, one can use the fact that $\frac{s^2}{\bar{x}}$ is distributed like χ^2 with n–1 degrees of freedom.

The theory behind the use of the ratio of variance to mean can be understood as follows: In a perfectly regular dispersion, the variation about the mean will be very small since all individuals are equally spaced, so you would expect most samples will have about the same number of individuals. Thus the low variance will result in a variance mean ratio less than one.

2.12 Constraints of Data Collection in Inland Open Waters

- Lack of clearly stated management objectives for fishery data collection.
- Lack of sufficient funding.
- Extreme variability (small-scale, part time and seasonal fishing operations).
- Inadequate system design and coordination (fragmented and incomplete data system).
- Inadequate information processing and analysis systems.
- Non-compliance with rules and regulations by the fishermen.
- Lack of skilled manpower.
- No involvement of local communities in management and data collection.
- Inadequacy of data collection systems.
- No sampling design.
- Reported data by fishermen leads to misreporting.
- Capture fishery mainly traditional—no fix time and landing centres.
- Grouping of data reported—no species data are available.
- Routine collection of biological data—only project based.
- Socio-economic data not collected on a routine basis.
- Frame survey—no information about fishing villages, fisher population and their gears and boats, landing centres, water bodies, etc.
- Lack of coordination between different agencies, institutes, states.
- No validation of data collected and compiled.
- Lack of two-way linkage among enumerators, data managers and end users; inappropriate data collection methods including insufficient use of sample survey.
- Non-standard classification and definition of water bodies, gears, boats, species, etc.
- Inaccurate reporting by administration.
- Discontinuity in funding disrupting time series data.
- Fishermen's unwillingness to give information.
- Lack of flexibility for reporting estimates at different levels.
- Cross-checking mechanism to validate survey estimate is not followed.
- Deliberate misreporting.
- Lack of attention to small-scale fishing activities.
- Lack of status, capacity or training of local fishery officers.

- Errors in catch reporting (often field data is collected based on 'recall'); difficulty in accessing sources of information (women, children and other fishers far from population centres).
- A reluctance to report catches because this is linked, in most countries, to licence fees or other forms of taxation.
- The collection of fish production statistics from inland water is beset with serious difficulties; the main in respect of riverine and estuarine fisheries is highly dispersed and isolated nature of fishing and landing areas.
- The migration of fishermen from one area to the other for fishing. The prevailing system of fish merchants buying off catches from the fishing boats at the fishing spots.
- The multispecies composition of catches and the landing of catches in unsorted condition.
- Isolated nature of fishing areas lower estuaries direct observations are very limited.

2.13 Conclusion

Sampling is the key to prudent inference on the population under consideration. Thus, the clear concept of sampling methods is essential for inland fisheries data analysis. This chapter provides the platform for concept building in sampling strategies. The sampling methods involve rather complicated mathematics, and mathematical derivation is avoided for the readers with little knowledge of mathematics. Implementation of the general theory of sampling is indeed a challenging task in real-life fisheries resource assessment. So, the chapter emphasizes the implementation part through some examples from inland fisheries, including fish sampling and water quality monitoring for different water bodies. Particular emphasis is given in ecological sampling, especially in the riverine ecosystem for which calibration of the general theory is of utmost importance. The examples provided will help the readers to implement scientifically designed monitoring with proper sampling techniques. The authors strongly recommended the readers to consult the suggested readings provided below for further knowledge gain.

Exercise

1. In a fishing village having 100 fishermen families, a survey was conducted to estimate the number of cows living in the community. A sample of 16 families was selected using simple random sampling for this study. The number of cows from each sampled family is shown below:
   0, 0, 0, 0, 0, 1, 1, 1, 1, 1, 1, 2, 2, 2, 3, 5.

   From this sample data, estimate the total number of cows living in the fishing village. Find the margin error and the confidence interval—assuming a 95% confidence level.

2. On a particular day, there were 60 coracles (a fishing unit) operating in a reservoir, and they came back with their catches to the same landing centre almost in a sequential order. A systematic sample of ten coracles was selected with a sampling interval of six units, and the catches (kg) from the sampled units were as follows:
   40, 50, 30, 60, 10, 40, 40, 10, 10, 20
   Estimate the total fish catch from the reservoir for the sampled day along with its standard error.
3. A reservoir has three fish landing centres categorized as low, medium and high, based on the volume of fish landings. The number of coracles (fishing units) operating in low, medium and high landing centres are 10, 60 and 40, respectively. A random sample of 10, 6 and 4 coracles was selected from low, medium, and high fish landing centres, respectively, and the catches (in kg) from the sampled coracles were recorded. The summary statistics are given in the following table.

Landing centre	Total no of coracles (Ni)	Total no of coracles sampled (n_i)	Average catch (xi)	Sample variance (s_i^2)
Low	100	10	19.5	25. 61
Medium	60	6	35.5	10.30
High	40	4	63.0	128.67

Estimate the total fish catch from the reservoir on the day of sampling with its standard error.

4. In a fishing village, there are 30 fishermen families, and they are grouped into 10 clusters—each cluster having 3 families. A random sample of 5 clusters was selected, and the monthly income (in Rs.) of each family from all selected cluster was collected. The details of the sample are given in the following table.

Cluster	Family 1	Family 2	Family 3
1	2750	3546	4333
2	3544	5426	5124
3	2765	1576	3675
4	3512	2765	2345
5	2675	4752	1894

(i) Estimate the average monthly income per fisherman family with its standard error.
(ii) Estimate the total income of all fisherman families in the village during the month. Give its standard error also.

5. List of all units of the population is called
   (a) Random sampling
   (b) Bias
   (c) Sampling frame
   (d) Probability sampling
6. Which one of these sampling methods is a probability method?
   (a) Quota
   (b) Judgement
   (c) Convenience
   (d) Simple random
7. The difference between a statistic and a parameter is called
   (a) Probability
   (b) Sampling error
   (c) Random
   (d) Non-random
8. Standard deviation of the sampling distribution of a statistic is called
   (a) Serious error
   (b) Dispersion
   (c) Standard error
   (d) Difference
9. Which of the following statement is true?
   (a) Standard error is always one.
   (b) Standard error is always zero.
   (c) Standard error is always negative.
   (d) Standard error is always positive.
10. Sampling error reduces by
    (a) Non-probability sampling
    (b) Increasing population
    (c) Decreasing the sampling size
    (d) Increasing the sampling size

Further Reading

Biradar RS (2002) Fisheries statistics, 2nd edn. Central Institute of Fisheries Education, Mumbai
Cochran WG (1977) Sampling techniques. Wiley Eastern Limited, New Delhi
Raj D, Chandhok P (1998) Sampling survey theory. Narosa Publishing House, New Delhi
FAO (2018) Fishery and aquaculture statistics 2016. FAO, UN
Kish L (1965) Survey sampling. Wiley, New York
Murthy MN (1977) Sampling theory and methods. Statistical Publishing Society, Calcutta
Sukhatme PV, Sukhatme BV, Sukhatme S, Ashok C (1984) Sampling theory of surveys with applications. Indian Society of Agricultural Statistics, New Delhi
Tyagi RK, Mandal SK (2008) Sampling methodologies for estimation of inland fish catch in India. CIFRI, Barrackpore, p 71

Descriptive Statistics for Inland Fisheries Data

3

Key Topics

- Concept of descriptive statistics
- Graphical presentation of data
- Importance of data processing
- Measure of central tendency

Statistics is basically a way of thinking about data that are variable. More specifically, statistics is a collection of methods for planning/designing studies or experiments, collecting/obtaining data and then organizing, presenting, analysing, interpreting and drawing conclusions. It is the science of learning from data. Statistical methods help in making rational and correct decisions in the face of variability and uncertainty. In fisheries science, where the information is both limited and varied, statistical tools are indispensable. To begin with, one should know about the data generated during the experiment/study, its distribution and its analysis to draw a valid conclusion from the experiment. Statistical methods can broadly be classified into two branches: descriptive statistics and inferential statistics. Descriptive statistics summarize a collection of data in a clear and understandable way by means of numerical (e.g. measures of central tendency and dispersion) and graphical (e.g. stem and leaf display, box plot, bar, histogram, etc.) approaches. Inferential statistics is generalizing from samples to populations using probabilities and includes testing of hypothesis (null hypothesis is put forward, and it is determined whether the data are strong enough to reject or accept it), point estimation (the sample is used to estimate a parameter) and interval estimation (confidence interval about the estimate is constructed). Descriptive and inferential statistics are interrelated. It is almost always necessary to use methods of descriptive statistics, to organize and summarize the information obtained from a sample before methods of inferential statistics can be used to make more thorough analysis of the subject

B. K. Das et al., *Concept Building in Fisheries Data Analysis*,
https://doi.org/10.1007/978-981-19-4411-6_3

under investigation. Furthermore, the preliminary descriptive analysis of a sample often reveals features that lead to the choice of the appropriate inferential method to be used later. Acquiring such skills forms an integral part of every discipline today. Generally, biologists keep themselves away from statistics, but they must know some basics of statistics that are essential for proper planning of experiments or studies, collecting, modelling/analysing data correctly and presenting/communicating their results appropriately. This chapter describes the basics of frequently used descriptive statistics.

3.1 Population and Sample

Population and sample are two basic concepts of statistics. Population can be characterized as the set of individual persons or objects in which an investigator is primarily interested during his or her research problem. Sometimes wanted measurements for all individuals in the population are obtained, but often only a set of individuals of that population are observed; such a set of individuals constitutes a sample. Thus, sample is that part of the population from which information is collected.

3.2 Parameter and Statistic

A parameter is a numerical quantity measuring some aspect of a population. For example, the mean is a measure of central tendency. A statistic is defined as a numerical quantity (e.g. the mean) calculated from a sample. Greek letters are used to designate parameters (see Table 3.1). Parameters are rarely known and are usually estimated by statistics computed from samples. For example, several parameters of great importance in statistical analyses and associated statistic used to estimate it from a sample are given below. However, there is no hard and fast rule for symbols used to denote parameter and statistic.

Table 3.1 Parameter and statistic

Quantity	Parameter	Statistic
Mean	μ	$\bar{x}$
Standard deviation	σ	S
Proportion	P	p
Correlation	ρ	r

3.3 Types of Data

1. Nominal data: Such type of data has only categories/names with no order or ranking. For example, fishes are categorized by sex as male or female; by species as rohu, catla, mrigal, etc.; and by group as Indian major carps, minor carps, exotic carps, featherbacks, etc.
2. Ordinal data: This type of data has categories that can be ordered or ranked. The differences between the ranks do not exist. For example, fish products ranked by consumers 1 = best, 2 = second best and so on.
3. Interval data: This type of data can be ranked and differences are meaningful. However, there is no meaningful zero, so ratios are meaningless, for example, temperature with the Celsius scale which has an arbitrarily defined zero point. Ratios are not allowed since 20 °C cannot be said to be "twice as hot" as 10 °C.
4. Ratio data: This type of data can be ranked, differences are meaningful, and there is a true zero. True ratios exist between the different units of measure, for example, weight of fish, length of fish, etc.

Sometimes, certain type of data may be converted from one form to another form for convenience. For example, length–weight of fishes is converted to their log values to obtain a linear relationship so that analysis becomes easy. Data can be transformed by taking the logarithm, square root, reciprocal or arcsine. If the dependent variable is nominal, ordinal, interval or ratio, then the following test are used: nominal, chi square and logistic regression; ordinal, chi square; and interval/ratio, correlation, multiple regression, path analysis, t-test, anova, manova and discriminant analysis.

3.4 Data Quality

The quality of data is of paramount importance and needs to be assessed carefully particularly for outliers, missing observations, etc. One should keep in mind that 'Garbage in, garbage out' is the rule of data processing. Thus, the first step after data collection is data editing and data cleaning which include detecting and correcting errors to ensure that the quality of the data is as high as possible. Any odd data should be carefully handled. If there are problems, the causes should be detected so that any outliers (extreme points) can be either rectified or discarded with adequate justification. Exploratory data analysis is very important as a single incorrect datum may result in skewness, hence need of repetition of the whole analysis. Therefore, actual statistical analysis should begin only when a researcher is fully confident that there are no errors or odd values.

3.5 Tabular and Graphical Presentation of Data

3.5.1 Frequency Distribution Table

It consists in dividing the range of values of the characteristic under consideration into different classes and counting the number, or the frequency, of observations within each class. The preparation of a frequency distribution from a list of the basic data consists of three steps: (a) specifying the classes into which data are to be grouped; (b) sorting the data into these classes; and (c) counting the number of observations in each class. It is easier to see patterns in the data, but the information about individual scores is lost. Graphs make it easier to see the trends in a set of data. Graphs for quantitative data include histogram, frequency polygon, etc., and that for qualitative data include bar chart, pie chart, etc.

Example 3.1 The length (in cm) of 30 randomly selected fish during harvesting from a floodplain wetland (beel) are as follows: 14, 25, 17, 20, 35, 38, 40, 25, 32, 31, 21, 27, 16, 11, 19, 22, 49, 30, 24, 29, 27, 34, 37, 35, 40, 26, 26, 34, 19 and 20.

Solution Here, maximum length is 49 and minimum length is 14. In order to have 8 class interval, the width of class interval will be $(49 - 14)/8 = 4$ approximately. The frequency distribution of the data is represented as in Table 3.2.

3.5.2 Scatter Diagram

A scatter diagram may be used for examining the relationship between two characteristics. For this diagrammatic presentation, a pair of observation is needed on each unit. For example, we wish to examine the relationship between the number of fishing units and daily total catch in a stretch of a river. Suppose 10 days of a month were selected following systematic sampling plan, and an investigator recorded details of the number of fishing boats and the total daily catch from the river stretch, as depicted in Table 3.3. The scatter diagram shown in Fig. 3.1 suggests the daily catch pattern.

Table 3.2 Frequency distribution of fish length data

Class interval	Tally marks	Frequency
10–14	//	2
15–19	////	4
20–24	~~////~~	5
25–29	~~////~~ //	7
30–34	~~////~~	5
35–39	////	4
40–44	//	2
45–49	/	1
Total		30

Table 3.3 Observations on the boats operating and daily catch in the river stretch

Day	1	2	3	4	5	6	7	8	9	10
No. of boat	14	17	12	14	20	16	8	17	18	11
Total catch (in kg)	490	720	330	540	800	560	140	550	750	370

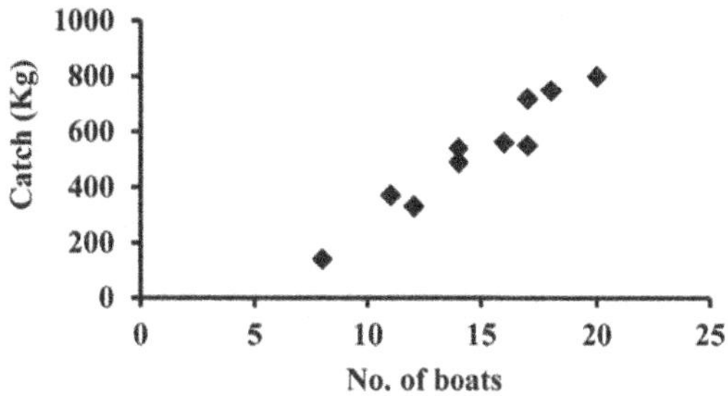

Fig. 3.1 Number of boats operating and total catch per day

Table 3.4 Consumers' preference for different fish species in Tripura

Species	Frequency
Rohu	67
Catia	42
Carpio	16
Hilsa	13
Mrigal	9
Other	3

Source: Das et al. (2013)

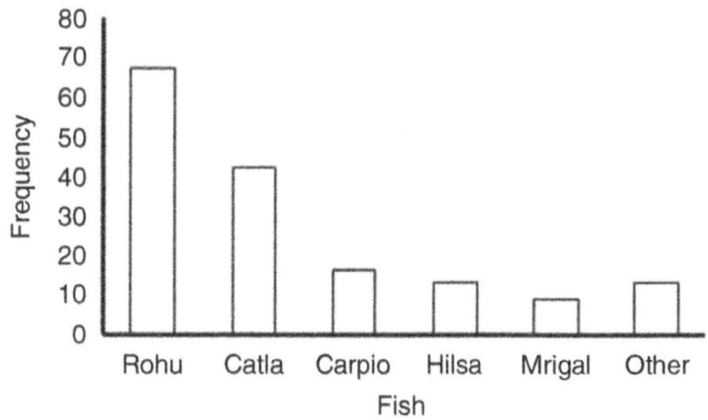

Fig. 3.2 Fish preference by consumers in Tripura

3.5.3 Bar Diagrams

It consists of lines or bars of equal width with length proportional to the value of variable or character under study. Bars may be drawn either horizontally or vertically. For example, data on consumer's preference for different fish species of Tripura is given in Table 3.4, and the corresponding bar diagram is shown in Fig. 3.2. When there are two or more comparable sets of closely related variables or characters, multiple bars are used. For example, data on fish production (in metric

Table 3.5 Fish production (MT) in beels under AFDC from 2010 to 2016

Year	Unstocked	Stocked	Total
2010–2011	941.4	1255.8	2197.2
2011–2012	616.3	1233.4	1849.7
2012–2013	1048.3	1656.9	2705.2
2013–2014	829.4	1507.8	2337.2
2014–2015	1520.5	2647.7	4168.2
2015–2016	1514.3	3106.0	4620.3
2016–2017	1606.7	3432.5	5039.2

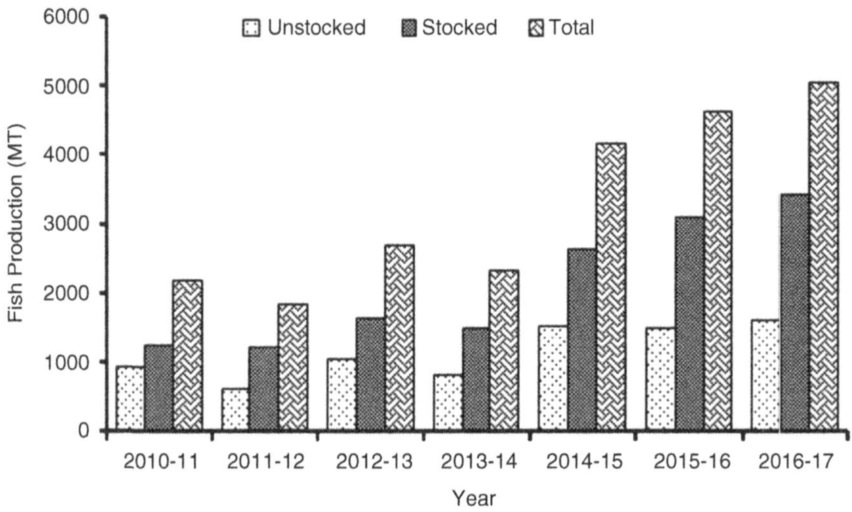

Fig. 3.3 Multiple bar diagram of fish production in unstocked, stocked and all beels of Assam

tons, MT) from stocked and unstocked floodplain wetlands (beels) under the administrative control of Assam Fisheries Development Corporation (AFDC) Ltd, Guwahati, from 2010–2011 to 2016–2017 are given in Table 3.5, and the corresponding multiple and component bar diagram in Figs. 3.3 and 3.4, respectively.

3.5.4 Pie Diagrams

A pie chart is a disk divided into pie-shaped pieces proportional to the relative frequencies of the classes. To obtain angle for any class, we multiply the relative frequencies by 360°, which corresponds to the complete circle. Either variables or attributes can be portrayed in this manner, but a pie chart is especially useful for attributes. A pie diagram for contribution of different fish groups/species to the total fish landings at a landing site of a river is shown in Fig. 3.5.

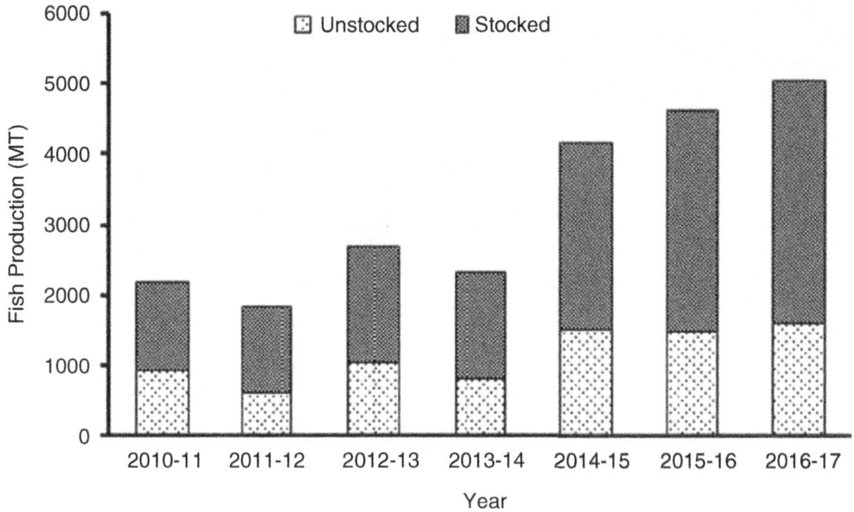

Fig. 3.4 Component bar diagram of fish production of stocked and unstocked beels in Assam

Fig. 3.5 Pie diagram for fish group composition by weight

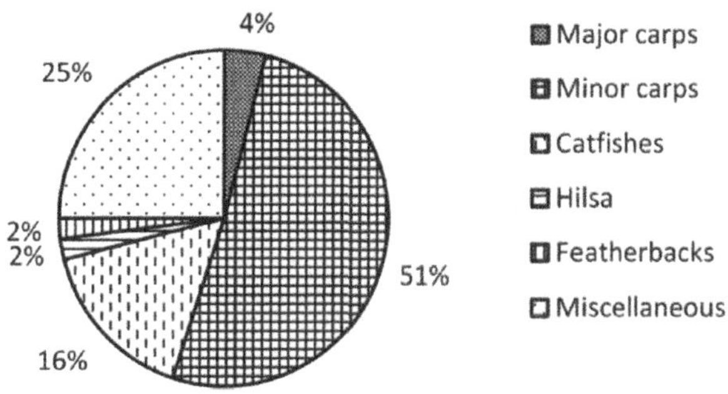

3.5.5 Histogram

Histogram is useful for studying the distribution of observed values of a variable.

A histogram is constructed from a frequency table. The intervals are shown on the x axis, and the number of scores in each interval is represented by the height of a rectangle located above the interval. This way we get a histogram corresponding to the data of class intervals of equal width. However, if the class intervals are of unequal width, the rectangles are drawn with heights proportional to the frequency density (ratio of class frequency to class width) so that the area is proportional to the class frequency. For constructing a frequency polygon, the middle points of the top side of the rectangles of a histogram are joined by a straight line. An additional class is taken at each end having zero frequency, and the polygonal line is closed by joining these points to the midpoints of adjacent rectangles. A frequency curve can be obtained by smoothing the frequency polygon. A histogram, frequency polygon

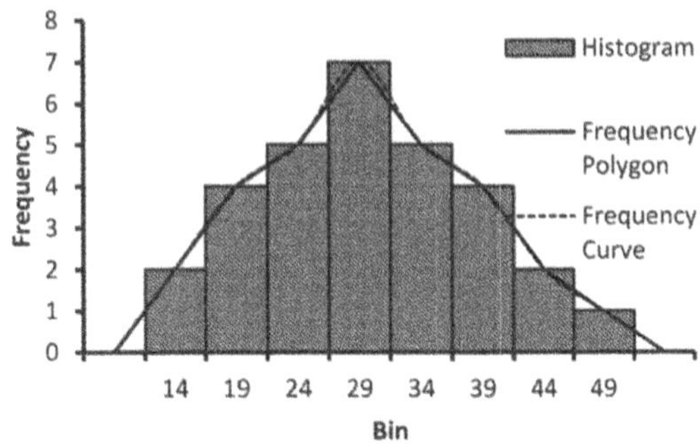

Fig. 3.6 Histogram, frequency polygon and frequency curve for fish length data set

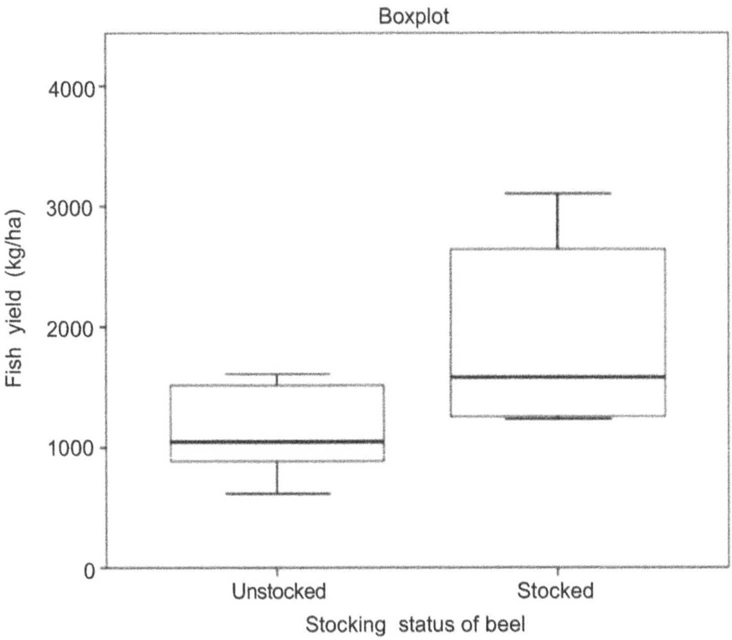

Fig. 3.7 Boxplot for fish yield in stocked and unstocked beels

and frequency curve of the frequency distribution of fish length obtained in Example 3.1 are given in Fig. 3.6.

3.5.6 Boxplot

Boxplot displays summary statistics about the distribution of observed values. It is a method to show the distribution of data both within and between groups. The boxplots (see Fig. 3.7) shows median and lower and upper quartiles and reveal whether the data is symmetric or skewed. It also helps to present the extreme values or commonly known as outliers, if present. The boxplot comparing fish yield versus stocking status of beel shows that stocking has an effect with respect to both location and variation. Stocked beels have high fish yield; variation in fish yield of unstocked beels is less in comparison to the stocked ones.

3.6 Measures of Central Tendency

Descriptive measures that indicate where the centre or the most typical value of the variable lies in collected set of measurements are called measures of centre. Measures of centre are often referred to as averages. The three most commonly used measures of central tendency are as follows: (1) mean (arithmetic, geometric and harmonic), (2) median and (3) mode. The mean and the median are used for quantitative data, whereas the mode can be applied with either quantitative or qualitative data.

3.6.1 Arithmetic Mean

The most commonly used measure of centre for quantitative variable is the (arithmetic) sample mean. It is defined as the average of data. The sample mean of the variable is the sum of observed values in a data divided by the number of observations. For example, if 5 observations of fish length (in cm) are 20, 38, 30, 38 and 21, then the arithmetic mean is given by $(20 + 38 + 30 + 38 + 21)/5 = 147/5 = 29.45$ cm. It is least affected by sampling fluctuations. It is based on all observations and affected by extreme values. The sum of deviations of the individual data elements from their mean is always equal to zero. In MS Excel, the mean is calculated using the formula '=AVERAGE (data range)'.

3.6.2 Median

The median is the middle value of the data. It is obtained by arranging data in order (either ascending or descending); the middle value is the median value. If there is no middle number, the median is the average of the middle numbers. It is used for scores and ranks both. For example, if 5 observations of fish length (in cm) are ordered as 20, 21, 30, 38 and 38, then the median is given by the centre value—at the third position in the example series, which is 30 cm. The median is not affected by extreme values and hence more representative than the mean in extremely skewed distributions. It is less reliable than mean as it takes into account only the number of observations and not their magnitude. In MS Excel, the median is calculated using the formula '=MEDIAN (data range)'.

3.6.3 Mode

The mode is defined as the value that occurs most in the data. It is obtained by arranging data in order (either ascending or descending) and then counting each one. The most frequently occurring value is the mode. The data with no repeats has no mode. Several repeats occurring the same number of times has multi modes. For example, if 5 observations of fish length (in cm) are ordered as 20, 21, 30, 38 and 38, here the value 38 occurs two times and is the mode. It is not based on all observations. Mode is useful in age and growth studies of fishes. For skewed distributions or samples where there is wide variation, mode and median are useful. In MS Excel, the mode is calculated using the formula '=MODE (data range)'.

The above definitions hold good for ungrouped data. For the same set of 5 observations of fish length, three different values, 29.4, 30 and 38 for the mean, median and mode, respectively, have been obtained. If the distribution is truly normal, the mean, median and mode coincide, i.e. all are equal to each other. The empirical relationship of mean, median and mode is given by mode $= 3 \times$ medium $-$ $2 \times$ mean.

3.6.4 Geometric Mean (GM)

If data are in geometric series (e.g. 2, 4, 8, 16… (x^2)), then geometric mean (GM) represents the central location. Geometric mean of the series is the nth root of product of observations. Log transformation is required before computing the GM. After computing the GM, it is then transformed back to the actual GM. GM can be obtained only if values assumed by the observation are positive. It is used when item increases with a compound rate, e.g. the growth of population, the growth of national income, bacterial culture growth, the growth of agricultural and industrial productivity, etc. For example, if the fish yield of a floodplain wetland (beel) after the adoption of supplementary stocking increases by 10% in the first year, 30% in the second year and 10% in the third year, the average increase per year is given by computing geometric mean, i.e. $(10 \times 30 \times 10)^{1/3} = 14.4$, whereas arithmetic mean is $(10 + 30 + 10)/3 = 16.7$. Note that for quantities such as rate of growth, GM is more appropriate than AM. In MS Excel, the geometric mean is calculated using the formula '=GEOMEAN (data range)'.

3.6.5 Harmonic Mean

Harmonic mean is the reciprocal of the arithmetic mean of reciprocals of values in the data set. It is used in the engineering field for computing average speed, in microbiology when concentrations are expressed as reciprocals, in post-harvest technology for the colour intensity of the product, and so on. For example, Rohu was purchased during the previous 2 weeks at 0.8 kg for Rs. 200 and 1 kg for Rs.200. The average purchase price for 1 kg is obtained by using harmonic mean, that is,

2/(0.8/200 + 1/200) = 222.2. Note here that for quantities involving two units like Rs/kg, HM is more appropriate than AM. The three means AM, GM and HM are related by AM $\geq$ GM $\geq$ HM. In MS Excel, the harmonic mean is calculated using the formula '=HARMEAN (data range)'.

3.6.6 Weighted Mean

Weighted mean is used when the relative importance of each observation is taken into account. The difference in importance (weight) may be due to a different number of observations for particular figures, or the values may carry different weights/ranks. It is obtained by dividing the sum of products of observed values and weights by the total weights. For example, the average fish production from floodplain wetlands (beels) of two districts of Assam is 24 tons and 13 tons. The averages were based on 1100 hectares and 750 hectares, respectively. The combined average yield per hectare is obtained by using weighted mean, i.e. $(24 \times 1100 + 13 \times 750) \div (1100 + 750) = 36150 \div 1850 = 19.54$ tons.

3.7 Measures of Dispersion or Variation

In addition to locating the centre of the observed values of the variable in the data, another important aspect of a descriptive study of the variable is numerically measuring the extent of variation around the centre. Two data sets of the same variable may exhibit similar positions of centre but may be remarkably different with respect to variability. Measures of variation are used mostly only for quantitative variables. The scatteredness of data from mean is called dispersion. Measure of the degree to which the observations vary about a central tendency is called the measure of dispersion. There are several different measures of variation. We will discuss most frequently used measures of variation: range, variance, standard deviation, standard error of mean and coefficient of variation.

3.7.1 Range

Range is the simplest measure of dispersion. The sample range of the variable is the difference between its maximum and minimum values in a data set. It is the simplest and a very crude measure of dispersion. It is not a good measure of variability as it is affected by extreme values and sample size. When expressing the range in writing, the lowest and highest observations or values themselves are shown rather than the calculated difference between them. For example, 'the average survival of *Labeo bata* fingerlings in Cage culture experiment was 85% (range 65–94%)'. In MS Excel, the range is calculated using the formula '=MAX (data range) – MIN (data range)'.

3.7.2 Variance and Standard Deviation (SD)

Variance (Var) is widely used and has become the fundamental basis for analysis as it is used for probability and hypothesis testing. The average of the squared deviations from the mean is variance. The standard deviation is the square root of the arithmetic mean of squares of deviation measured from the mean. In other words, it is the positive square root of variance. It is less affected by fluctuation of sampling and its value is always definite. Variance is the most effective measure of variability. It is more appropriate only if data under consideration is normally distributed. If individual observations are clustered and evenly scattered around the sample mean, it helps to calculate a range that will include a given percentage of observation. For example, if $N \geq 30$, the range Mean ± SD includes 68% of observations, Mean ± 2 (SD) includes 95% of observations, and the range Mean ± 3 (SD) includes 99% of observation. In MS Excel, the variance of population and sample is calculated using the formula '=VAR.P(data range)' and '=VAR.S(sample range)', respectively. Similarly, SD of population and sample is calculated using the formula '=STDEV.P (data range)' and '=STDEV.S (sample data range)', respectively.

3.7.3 Standard Error

Standard error (SE) has become popular in the last decade. Researchers often misunderstand and misuse it. Variability of observations within a data set is SD, whereas variability of two or more means is SE. Therefore, it is often called standard error of means. It is computed as SD of two or more means divided by the square root of the number of observations: $SE = \frac{SD}{\sqrt{n}}$. The standard error of a statistic is the standard deviation of the sampling distribution of that statistic. Standard errors are important because they reflect how much sampling fluctuation a statistic will show.

The inferential statistics involved in the construction of confidence intervals and significance testing are based on standard errors. The standard error of a statistic depends on the sample size. In general, the larger the sample sizes, the smaller the standard error.

From a statistical point of view, presenting means without variability (SD or SE) has no meaning. Either SD or SE needs to be presented along with the means while presenting results both in tabular and graphical forms. More importantly, error bars drawn using SE or SD may serve as rough indicators of the presence or absence of significant differences between means. However, this has to be confirmed by using the appropriate statistical tool. There is no direct function for computing SE of mean in MS Excel. However, it can be calculated by using formula '=STDEV.S(data range for means)/SQRT(*n*)'.

3.7.4 Coefficient of Variation (CV)

It is a relative measure of variation and widely used to compare two or more statistical series. The statistical series may differ from one another with respect to their mean or standard deviation or both. Sometimes they may also differ with respect to their units, and then their comparison is not possible. To have a comparable idea about the variability present in them, CV(%) is used. It is defined as the ratio of standard deviation to arithmetic mean and is expressed in percentage. It is a unit-free measure and always expressed as percentage. It is useful in comparing the reliability of results obtained in different experiments or investigations. Of the two groups, the one with less CV is said to be more consistent. It is informative if it is given along with the mean and standard deviation. Otherwise, it may be misleading.

For example, length and weight were recorded on a random sample of 100 fishes of a particular species. It was found that mean length was 50 cm with standard deviation of 20 cm, whereas the mean weight was 200 g with standard deviation of 65 g. Naturally, comparison of weight and length variation is not meaningful, as units of measurements are different. Therefore, CV for length is $(20/50) \times 100 = 40\%$, and CV for weight is $(65/200) \times 100 = 32.5\ \%$. Thus, the length is more variable than the weight since the CV of length is more than the CV of weight.

3.7.5 Shape of Data Set/Distribution

A distribution is said to be skewed if one of its tails is longer than the other. The first distribution shown in Fig. 3.8 is a positive skew since it has a long tail in the positive direction. The second distribution beside it has a negative skew since it has a long tail in the negative direction. Finally, the last distribution is symmetric and has no skewness. Distributions with positive skew are sometimes called 'skewed to the right', whereas distributions with negative skew are called 'skewed to the left'. Here, we are not going in mathematical details of computing these two parameters as these can obtained using any statistical software including MS Excel. The mean is larger than the median in positively skewed distributions, whereas it is smaller than the median in positively skewed distributions.

On the other hand, kurtosis is based on the size of a distribution's tails. Distributions with relatively large tails are called 'leptokurtic'; those with small tails are called 'platykurtic'. A distribution with the same kurtosis as the normal

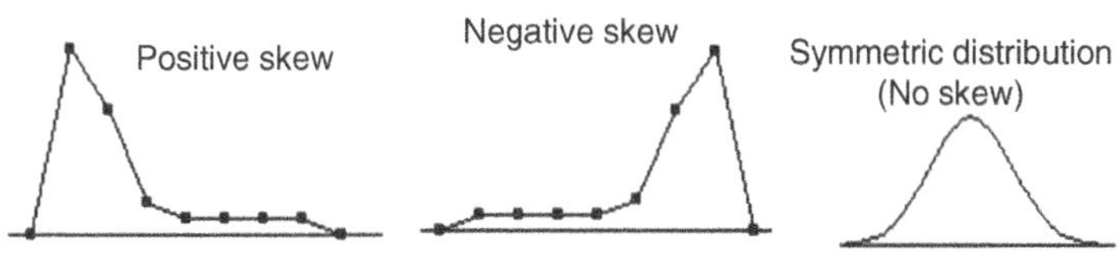

Fig. 3.8 Different types of skewed distributions

Table 3.6 Threshold values for categorizing skewness and kurtosis of a distribution

Skewness	<−1	0	>+1
	Left-skewed	Normal	Right skewed
Kurtosis	<2	3	>4
	Platykurtic	Normal or mesokurtic	Leptokurtic

distribution is called 'mesokurtic'. Kurtosis with zero value indicates a symmetrical distribution. Normal distributions produce a kurtosis statistic of about zero. A summary of skewness and kurtosis in relation to normality is given in Table 3.6.

3.7.6 Types of Distribution

The data tends to follow normal distribution as sample size increases indefinitely which is the key feature of inferential statistics. There are two types of distribution: normal distribution and non-normal distribution. In normal distribution, the data is perfectly symmetrically distributed about mean and forms a bell-shaped curve in frequency distribution plot. The height of a normal distribution can be specified mathematically in terms of two parameters: the mean (μ) and the standard deviation (σ). The standard normal distribution is a normal distribution with a mean of 0 and a standard deviation of 1. Normal distributions can be transformed to standard normal distributions by the formula: $Z = \frac{(X-\mu)}{\sigma}$, where X is a score from the original normal distribution, and μ and σ are the mean and standard deviation, respectively, of the original normal distribution. One important use of the standard normal distribution is for converting between scores from a normal distribution and percentile ranks. Typically, a researcher is interested in how well the distribution can be approximated by the normal distribution. Simply by putting the raw data directly into a software and applying normality tests (e.g. Kolmogorov-Smirnov test or Shapiro-Wilk test) can help in judging whether our data is normally distributed or not. This can also be done by visual examination of data using a histogram with normal curve superimposed over it. All distributions in universe tend to follow normal distribution as sample size increases. Various inferential methods like *t-test* (used for testing single mean, two group means and paired observations) and ANOVA (used for testing more than two group means) are based on the assumptions of normality. The distribution is said to be non-normal if data is skewed on one side. A distribution curve is said to be positively skewed if (mean-median) > 0, otherwise negatively skewed if less than 0. For example, binomial distribution in the event can have one of the two possible outcomes such as survival/death, success/failure, etc.

3.8 Conclusion

This chapter describes some of the basic statistics and their application in fisheries, focusing on types of data, collection, organization and summarization of data along with some discussion made on type of distributions and type of tests to be used for different types of data. This will help the readers to have a preliminary study of your data and form a basis, before proceeding to inferential statistics. This chapter is not an exhaustive coverage of basic statistics relevant to fisheries science. For further study, the readers can consult the suggested readings listed at the end. However, all these preliminary studies of data can be carried out with the help of computers and statistical software like SPSS, Statistica, MS Excel, SAS, R, etc. Among all the statistical software, MS Excel is widely used because of easiness and wide coverage.

Exercise

1. Classification of respondents only on the basis of gender is an application of
   (a) Nominal scale
   (b) Ordinal scale
   (c) Interval scale
   (d) Ratio scale
2. The median of fish length (cm) 70, 64, 80, 74, 92 is
   (a) 76
   (b) 74
   (c) 77
   (d) 80
3. The relationship between AM, GM and HM is
   (a) AM $\leq$ M $\leq$ HM
   (b) AM $\leq$ HM $\leq$ GM
   (c) M = HM = GM
   (d) AM $\geq$ GM $\geq$ HM
4. For nominal data, the meaningful measure of the central tendency is
   (a) Arithmetic mean
   (b) Median
   (c) Mode
   (d) Both A and B
5. .... cannot be easily approximated from a boxplot.
   (a) Range
   (b) Median
   (c) Standard deviation
   (d) Interquartile range
6. A distribution is said to be positively skewed if
   (a) Mean >Median
   (b) Mean = Median

   (c) Mean <Median
   (d) None of these
7. Which of the following is a relative measure of dispersion?
   (a) Standard deviation
   (b) Coefficient of variation
   (c) Variance
   (d) Coefficient of kurtosis
8. The range of the scores 62, 196, 123, 64, 139, 80 is
   (a) 134
   (b) 18
   (c) 196
   (d) 142
9. The variance of fish length (cm) 19, 21, 23, 25 and 27 is 8. The variance of 38, 42, 46, 50 and 54 is
   (a) 8
   (b) 6
   (c) 16
   (d) 32
10. Geometric mean of 2, 4, 8 is
   (a) 4
   (b) 2
   (c) 8
   (d) 5
11. Represent the following data on marine and inland fish production (in 000′ tons) from 2011–2012 to 2016–2017 by (1) multiple bar diagram and (2) component bar diagram.

Year	2011–2012	2012–2013	2013–2014	2014–2015	2015–2016	2016–2017
Inland	5294	5720	6132	6509	7213	7769
Marine	3372	3321	3440	3655	3583	3641

12. The following data shows consumption expenditure (in rupees) of a fisher family before and after adoption of pen culture technology. Represent both before and after expenditure by using pie diagram.

	Items/heads	Expenditure (in rupees)
	Before	After
Clothing	1600	3000
Food	2400	3800
Education	1000	2000
Electricity	1000	1500
Others	2000	1700

13. Calculate arithmetic mean, median and mode for the following data on fish length (cm): 18, 26, 23, 17, 23, 21, 23, 22, 38, 23. Which are the correct values?
    (a) Mean = 23, median = 17, mode = 22
    (b) Mean = 23, median = 22 and mode =23
    (c) Mean = 18, median = 23 and mode = 23
    (d) Mean = 22, median = 23 and mode = 23
14. If mode and median of data on length of randomly selected fishes are 64 and 48, respectively, obtain arithmetic mean of fish length.
    (a) 40
    (b) 41
    (c) 60
    (d) 30
15. The monthly salary (in lakh rupees) and the number of investigators involved in a research project are given below.

	Monthly salary (Rs.)	Number
Principal scientist	2.5	2
Senior scientist	2.0	6
Scientist	0.8	12

What is the average monthly salary?
(a) 3.00 lakhs
(b) 1.33 lakhs
(c) 2.00 lakhs
(d) 1.70 lakhs

16. HDPE net used in GI cage technology is assumed to depreciate 40% in value in the first year, 25% in the second year and 10% per annum for the next 3 years, each percentage value being calculated on the diminishing value. What is the average percentage depreciation for the 5 years?
    (a) 30%
    (b) 35%
    (c) 20%
    (d) None of the above
17. A boat covers a distance from station A to station B at 40 km/h and returns by the same route at 60 km/h. What is the average speed for the whole journey?
    (a) 41 km/h
    (b) 40 km/h
    (c) 48 km/h
    (d) None of the above
18. Compute range, variance, standard deviation (SD) and coefficient of variation (CV) for the following data on fish weight (gm), 62, 196, 123, 64, 139, 80, 123, 132, 138, 139, and choose the correct answer from the following four options.
    (a) Range = 134, variance = 1675.82, SD = 40.94, CV = 34.23%
    (b) Range = 131, variance = 1575.82, SD = 45.94, CV = 24.23%

(c) Range = 124, variance = 1675.82, SD = 30.94, CV = 24.23%
(d) None of the above

19. Which measure of central tendency will be appropriate for (1) fish yield of floodplain wetlands in two districts; (2) size of agricultural holdings; (3) rating of a fish recipe by respondents; and (4) fish length (cm) 12, 10, 14, 6, 9, 13, and X (X < 7)?
    (a) (i) Median, (ii) mean, (iii) weighted mean, (iv) geometric mean
    (b) (i) Mean, (ii) median, (iii) mode, (iv) geometric mean
    (c) (i) Mean, (ii) mode, (iii) median, (iv) median
    (d) None of the above
20. The data on the number of fish caught by the two nets in a fishing season is given below.

Number of fishes caught					
	0	1	2	3	≥4
Net A	24	15	14	7	8
Net B	20	18	24	15	12

Find which net may be considered more efficient in its performance?

(a) Net A is more consistent.
(b) Net B is more consistent.
(c) Consistency of both nets is the same.
(d) None of the above.

Further Reading

Anderson TW, Sclove SL (1974) Introductory statistical analysis. Houghton Mifflin Company, Boston

Bhujel RC (2008) Statistics for aquaculture, 1st edn. Wiley, Hoboken

Biradar RS (2002) Fisheries statistics, 2nd edn. Central Institute of Fisheries Education, Mumbai

Clarke GM, Cooke D (1998) A basic course in statistics. Arnold, London

Das A et al (2013) Fish consumers' behaviour at selected fish markets of Tripura, India. Fish Technol 50(2):185–190

Roy AK, Sarangi N (2008) Applied bioinformatics, statistics & economics in fisheries research. New India Publishing Agency, New India

Basic Concept of Hypothesis Testing and Parametric Test

4

Key Topics
- Importance of testing of hypothesis in fisheries
- Concept of hypothesis testing for inference
- One sample parametric test of significance
- Two sample parametric test of significance

Statistical analysis is done by many ways, but the majority of biologists use the 'classical' statistics which involves testing of null hypothesis using experimental data. In this process we estimate the probability that is obtained from the observed results or, something more extreme, if the null hypothesis is true. If the estimated probability (the *p*-value) is lesser than the significance value, we conclude that null hypothesis is unlikely true, and we reject the null hypothesis. So, hypothesis is defined as an assumption about a single population or about the relationship between two or more populations. It is testable and provides a possible explanation of a certain phenomenon or event. On the other hand, if a hypothesis is not testable, then it implies insufficient evidence to provide more than a tentative explanation, e.g. extinction of inland fishes. To test the new information or knowledge or belief about the populations against the existing one, two hypotheses are used, the null hypothesis (H_0) and the alternative hypothesis (H_1). Null hypothesis (H_0) assumes that there is no difference between the new and existing populations. However there can be some indications that existing knowledge of beliefs may not be true. The null hypothesis (H_0) is tested against the alternative hypothesis (H_1). The alternative hypothesis states the statistical statement indicating the presence of an effect or a difference and sometimes known as 'an intelligent guess' based on limited information. It may so happen that the experimental results may match predictions, but it should be believed after proper testing and analysis of appropriate statistical tool(s), which includes designing an experiment or survey to generate or collect data/

B. K. Das et al., *Concept Building in Fisheries Data Analysis*,
https://doi.org/10.1007/978-981-19-4411-6_4

information (raw), exploratory data analysis and choosing an appropriate significance level or confidence limits (intervals).

In the process of hypothesis testing, only null hypothesis is tested; based on the data outputs or results, it is either accepted or rejected, which is expressed as follows:

$$H_0 = H_1 \text{ which means there is no difference} \rightarrow \text{Accept}$$

$$H_0 \# H_1 \text{ which means there is a difference} \rightarrow \text{Reject}$$

As the hypothesis is either accepted or rejected based on the numerical facts, only trustworthy data can be used for this purpose. Hypothesis testing therefore involves data collection using proper methods, compilation and securitization, use of appropriate tools for analysis and judicious decision, interpretation and explanation.

For an example, suppose a researcher is interested to conduct a study in which n samples of fishes are randomly selected from the fish population size N and their weight is measured. Now he/she is interested to test the mean weight of the fishes if 1000 g or not. The null and alternative hypothesis can be written as H_0: $\mu = 1000$ g and H_1: $\mu \neq 1000$ g. Here null hypothesis shows that the mean weight of the population represented by the samples n is 1000 g, and the alternative hypothesis is that the mean of the population represented by the sample is not equal to 1000 g. In the above situation, the alternative hypothesis is non-directional, and it does not make a statement towards specific direction. The alternative hypothesis H_1: $\mu \neq$ 1000 g states only the fish population mean is not equal to 1000 g, but it does not state whether it will be less or greater than 1000 g. The directional alternative hypothesis can be represented in two directions (referred as one-tailed) H_1: $\mu >$ 1000 g or H_1: $\mu <$ 1000 g. The alternative hypothesis H_1: $\mu >$ 1000 g states that the mean of the fish population represented by sample value is greater than 1000 g. If the directional alternative hypothesis H_1: $\mu >$ 1000 g is employed, the null hypothesis can only be rejected if the data indicate that fish population mean is some value above 1000 g. The alternative hypothesis H_1: $\mu <$ 1000 g states that

Note: In general, researcher selects the alternative hypothesis as non-directional when there is no expectation of the proposed research output and chooses directional alternative hypothesis if one have a definite expectation about output of the experiment. For non-directional alternative hypothesis, large effect or difference in sample data is obtained as compared to the directional alternative hypothesis.

4.1 Significance Level

The mean of the fish population represented by sample value is less than 1000 g. If the directional alternative hypothesis H_1: $\mu <$ 1000 g is employed, the null hypothesis can only be rejected if the data indicate that fish population mean is some value less than 1000 g.

The term significance implies that the obtained results from an experiment (s) differ from the hypothesis due to chance or in reality. The decision on null

Table 4.1 Critical values for Z-test at 0.05 and 0.01 significance level

	$z_{0.05}$	$z_{0.01}$
Two-tailed values	1.96	2.58
One-tailed values	1.65	2.33

hypothesis either to accept or reject is based on contrasting the observed outcome of an experiment with the outcome one can expect if, in fact, the null hypothesis is true. This decision is made by using the appropriate inferential statistical test. An inferential statistical test is designated as a test statistic which is evaluated in reference to a sampling distribution, which is a theoretical probability distribution of all the possible values the test statistic can assume if one were to conduct an infinite number of studies employing a sample size equal to that used in the study being evaluated. The probabilities in a sampling distribution are based on assumption that samples are randomly drawn from the representing population. Researcher's decision to declare the difference is statistically significant is based on an analysis of sampling distribution of the test statistic. If the test statistic is found to be significant at the given level of significance, then the result is due to the effect of treatment in the experiment, not a chance. In relation to this, the most likely decision is to conclude that the difference is due to the experimental treatment and thus reject the null hypothesis. In general for scientific purpose, there should not be more than a 5% likelihood that the difference is due to chance. If one thinks that 5% is too high, one may choose to employ 1%, or an even lower minimum likelihood, before concluding that a difference is significant. If the $p > 0.05$ is observed which indicates that the result of an experiment is not significant. This notation indicates that there is greater than 5% likelihood that an observed difference or effect could be due to chance. On the other hand, the notation $p < 0.05$ indicates that the outcome of a study is significant at the 5% level. This indicates that there is less than a 5% likelihood that an obtained difference or effect can be due to chance. The notation $p < 0.01$ indicates a significant result at the 0.01 level (i.e. there is less than a 1% likelihood that the difference is due to chance). The value $Z = 1.96$ is referred to as the tabulated critical two-tailed 0.05 Z value. This value is employed since the total proportion of cases in the normal distribution that falls above $Z = +1.96$ or below $Z = -1.96$ is 0.05. The value $Z = 2.58$ is referred to as the tabled critical two-tailed Z value. This value is employed since the total proportion of cases in the normal distribution that falls above $Z = +2.58$ or below $Z = -2.58$ is 0.01. The value $Z = 1.65$ is referred to as the tabled critical one-tailed 0.05 Z value. This value is employed since the proportion of cases in the normal distribution that falls above $Z = +1.65$ or below $Z = -1.65$ in each tail of the distribution is 0.05. The value $Z = 2.33$ is referred to as the tabled critical one-tailed 0.01 Z value. This value is employed since the proportion of cases in the normal distribution that falls above $Z = +2.33$ or below $Z = -2.33$ in each tail of the distribution is 0.01. The probability value which identifies the level of significance is represented by the notation α which is the lowercase Greek letter alpha (Table 4.1).

4.2 Type of Errors

To design a research trial, a researcher has to make proper design or survey and to select suitable statistical tools while analysing data. Minimization of errors plays a crucial role. However, in reality errors occur quite often. In the framework of hypothesis testing, researcher commits two types of errors, Type I and Type II error. In Type I error, the null hypothesis is rejected when it is true (i.e. it is considered that false alternative hypothesis is true). The likelihood of committing a Type I error is specified by the α level when a researcher employs in evaluating an experiment. A Type II error occurs when a false null hypothesis is accepted (i.e. it is considered that a true alternative hypothesis is false). The likelihood of committing a Type II error is represented by β, which is the lowercase Greek letter beta. The likelihood of rejecting a false null hypothesis represents what is known as the power of a statistical test. Committing Type II error is more harmful than committing Type I error. So the test procedure is formulated in such a way that Type II error is minimized with fixed level of Type I error. This fixed level of Type I error is knows as level of significance (α). Table 4.2 shows the summary of the errors.

4.3 Power of Test

The power of a test is determined by subtracting the value of beta from 1 (i.e. power $= 1 - \beta$). The likelihood of committing a Type II error is inversely related to the likelihood of committing a Type I error. In other words, as the likelihood of committing one type of error decreases, the likelihood of committing the other type of error increases. Thus, with respect to the alternative hypothesis one employs, there is a higher likelihood of committing a Type II error when alpha is set equal to 0.01 than when it is set equal to 0.05. The likelihood of committing a Type II error is also inversely related to the power of a statistical test. In other words, as the likelihood of committing a Type II error decreases, the power of the test increases. Consequently, the higher the alpha value (i.e. the higher the likelihood of committing a Type I error), the more powerful the test. In other words, the power of the test is the probability of making the right decision or detecting significant difference when it exists. Therefore, power analysis is important, especially when the results are non-significant. Power analysis can reveal whether the replication was adequate for any treatment to show its effects. A minimum of 80% (or $\beta = 0.20$) is considered acceptable statistical power (Searcy-Bernal 1994). The higher the value of power, the better the test.

Table 4.2 Types of errors in hypothesis testing

Decision	When H_0 is true	When H_0 is false
H_0 rejected	Type 1 error (α)	√
H_0 accepted	√	Type II error (β)

Example 4.1: Illustration of type I and type II errors The concept of Type I and Type II errors is illustrated here by a case study. An experimental trial was conducted to test the tilapia production of a new strain of tilapia in comparison to the local tilapia strain. A researcher states the null hypothesis and alternative hypothesis as H_0: Local tilapia strain production = new tilapia strain production and as H_1: new tilapia strain production >Local tilapia strain production. Now researcher 'A' (let) found that a new strain of tilapia gave higher production compared to the local strain (select alternative H_1 hypothesis) and recommended the farmers to grow a new one. But, researcher 'B' later used the same experimental protocols and did not actually find any statistical difference. In this situation researcher 'A' made a Type I error. In another trial with catfish, researcher 'A' did not see any difference between new and local strain and select null hypothesis that production of both strains is the same. But, later researcher 'B' found new strain could actually produce more and statistical difference was detected. Now researcher 'A' made a Type II error. Now by comparing the Type I error with Type II error, it was observed that Type I is more harmful than the Type II since in the case of tilapia, many farmers might have spent a lot to purchase the new strain as per recommendation to replace the existing stock and also change the existing practices and facilities, whereas in the case of catfish, farmers didn't need to change anything—they just followed their existing protocols. This means that there was no additional cost involved. Although both of these errors are unwanted and should be avoided, the case of tilapia (Type I error) is more dangerous than the case of catfish (Type II error).

4.4 Confidence Level, Limits and Interval

The decision of hypothesis that either it is true or false is taken at a certain level of confidence. In statistical point of view, nothing is absolutely true or false that does not support 100% confidence. In social survey research, even 90% confidence level is enough, but in biological research 95% is considered as sufficient, but in medical research, confidence level is usually as high as 99%. Any mean (parameter in broad sense) has two confidence limits for the given level of confidence, i.e. lower confidence limit (LL) and the upper confidence limit (UL). The difference between these two is known as confidence interval (CI). The sample mean estimates the true mean, and standard error (SE) describes the variability of that estimation. The variability is expressed in terms of probabilities by calculating CIs.

Statistically, a confidence interval estimate of a parameter consists of an interval of numbers along with a probability that the interval contains the unknown parameter, and the level of confidence is a probability that represents the percentage of intervals that will contain if a large number of repeated samples are obtained. A 95% level of confidence means that if 100 confidence intervals were constructed, each based on a different sample from the same population, we would expect 95 of the intervals to contain the population mean. The three factors are needed to construct a confidence interval for the population mean (a) point estimate of population, (b) level of confidence and (c) standard deviation of the sample mean.

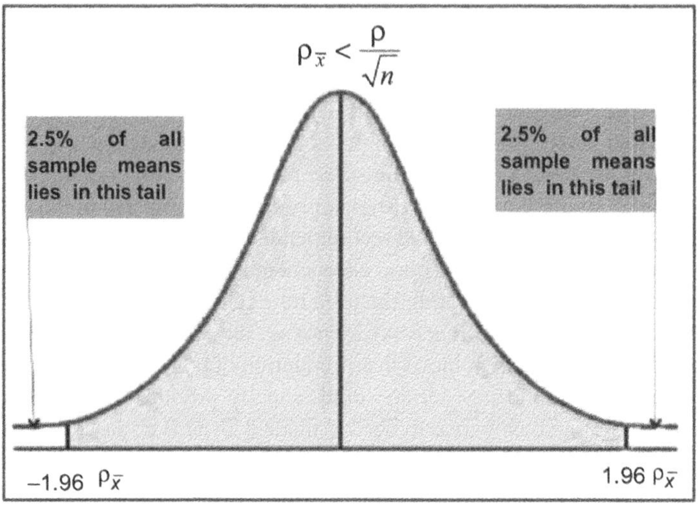

Fig. 4.1 Normal plot with 95% confidence limit between $-1.96\ \sigma_{\overline{x}}$ and $1.96\ \sigma_{\overline{x}}$

Suppose we obtain a simple random sample from a population. Provided that the population is normally distributed or the sample size is large, the distribution of the sample mean will be normal with mean (μ) and standard deviation ($\sigma/\sqrt{n}$), since here is normally distributed, we know that 95% sample means should lie within 1.96 standard deviations of the population mean μ and 2.5% will lie in each tail (see Fig. 4.1). Of all sample means, 95% are in the interval

$$\mu - 1.96\frac{\sigma}{\sqrt{n}} < \overline{x} < \mu + 1.96\frac{\sigma}{\sqrt{n}} \tag{4.1}$$

and the probability of this interval is

$$P\left(\mu - 1.96\frac{\sigma}{\sqrt{n}} < \overline{x} < \mu + 1.96\frac{\sigma}{\sqrt{n}}\right) = 0.95 \tag{4.2}$$

4.4.1 Interpretation of Confidence Interval

A $(1-\alpha) \times 100\%$ confidence interval means that if we obtain many simple random samples of size μ from the population mean, μ, which is unknown, then approximately $(1-\alpha) \times 100\%$ of the intervals will contain μ. For constructing a $(1-\alpha) \times 100\%$ confidence interval about l and o known, then the confidence limit is given by lower confidence limit (CL) as $\overline{x} - Z_{\alpha/2} \cdot \frac{\sigma}{\sqrt{n}}$ and $\overline{x} + Z_{\alpha/2} \times \frac{\sigma}{\sqrt{n}}$ upper limit (see Fig. 4.2). Here samples are drawn from the normal population.

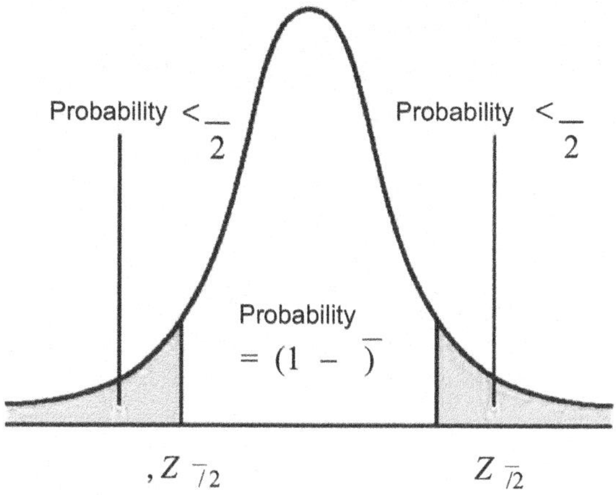

Fig. 4.2 A generic $(1-\alpha) \times 100\%$ confidence interval for the mean of a normally distributed population

Example 4.2 The following example shows that when the population variance is unknown, then using student's t-statistic, we can compute 68%, 95% and 99% CIs. Suppose a sample of 40 fish was drawn from a pond containing 2,000 fish. If the computed mean is 500 g and standard deviation (SD) is 33 g, what are the LL and UL for 68%, 95% and 99% CIs?

Solution *For this, SE is computed as*

$$SE = \frac{SD}{\sqrt{n}} = 5.16\,\text{g}$$

The range mean $\pm 1 \times SE$ *covers 68% in the normal curve. Or by simply adding/subtracting* $1 \times SE$ *to/from the mean, we can get the LL and UL for 68% CI:*

$$\begin{aligned} LL &= 500 - 1.16 \\ &= 497.84\,\text{g} \\ UL &= 500 + 5.16 \\ &= 505.16\,\text{g} \\ CI &= 505.16 - 494.84 \\ &= 10.32\,\text{g} \end{aligned}$$

To calculate the 95% and 99% CIs, SE has to be multiplied by t-statistics, which depend on the degree of freedom (df), before adding to or subtracting from the mean (the critical value for 39 df of the t-distribution): t-statistics for 95% CI $= t_{(0.05,39)} \times$ SE for $n^{-1} = 39$; *i.e.* $2.023 \times 5.16 = 10.43$ g.

Thus,

$$
\begin{aligned}
L &= 500 - 10.43 \\
&= 489.46 \text{ g} \\
UL &= 500 + 10.43 \\
&= 510.43 \text{ g} \\
CI &= 510.43 - 489.56 \\
&= 20.87 \text{ g}
\end{aligned}
$$

*In this case, we can now say with 95% confidence that the true mean falls between these limits (*489.56–510.43*). Similarly, CI further increases as we want higher confidence, e.g. for 99%: t-statistics = mean* ± (2.708 × 5.16) = 500 ± 13.97 g,

$$
\begin{aligned}
LL &= 500 - 13.973 \\
&= 486.36 \text{ g} \\
UL &= 500 + 13.97 \\
&= 513.97 \text{ g} \\
CI &= 513.97 - 486.36 \\
&= 27.94 \text{ g}
\end{aligned}
$$

This above example shows that as *CI* increases, the confident level also increases. Wider range is necessary to be confident that the sample mean will fall within the range.

Statistical Significance and Biological Significance

Let us assume if a feed item vitamin C at 50 g kg^{-1} of feed is given to the fish and the survival is increased by 10%. This 10% survival that is claimed due to the use of vitamin C is substantial and has an economic value for farmers. However, an increase lower than this may be still statistically significant but may not be considered substantial.

4.5 Selection of Statistical Tools

The appropriate selection of statistical tools is necessary after carefully examining the conditions because even if the data have been generated perfectly and collected carefully, the misuse of statistical tools might result in wrong conclusions and provide bad recommendation. Researchers need to determine which distribution a particular data set follows: normal, binomial, Poisson or free of any distribution. The fundamental principle is that, if the data sets are normally distributed sample means, standard deviations represent the population but, if the data sets are distribution-free, then these parameters do not represent the population. Therefore, use of the mean values to represent and compare between/among them has no meaning. In such cases, non-parametric tests are used. These tests do not consider the actual figures and degree of deviations from central tendency but simply use ranks assigned to the data points. Whatever may be the tools selected for testing of hypothesis, the following objectives are set to perform any statistical test of significance:

- State the null and alternative hypotheses.
- Choose a level of significance and sample size *n*.
- Determine the appropriate test statistic and sampling distribution.
- Determine the critical values that divide the rejection and non-rejection regions.
- Collect data and compute the value of the test statistic.
- Make the statistical decision and state the conclusion.

4.5.1 One Sample Parametric Test

A test of significance allows us to draw conclusions or make decisions regarding population from sample data. When sample is drawn from single population to make inference on the population, we call it as a signal sample problem. Usually a population is characterized by some parameter. There are several parametric tests by which inference is made on the population parameter based on a random sample. In this section, we describe the most popular and widely used parametric test applied in one sample problem.

4.5.2 Large Sample Test of Significance for Population Parameter

For large (sample size), almost all the distributions can be approximated very closely by a normal distribution. Therefore the **normal test of significance** is used for large samples. If u is any statistic (function of sample values), then for large sample,

$$Z = \frac{u - E(u)}{\sqrt{V(u)}} \sim N(0, 1)$$

where $E(u)$ = expectation of u and $V(u)$ = variance of u and $N(0, 1)$ = standard normal distribution, i.e. normal distribution with 0 mean and variance 1.

Thus if the discrepancy between the observed and the expected (hypothetical) values of a statistic is greater than Z_α times the standard error (SE), the hypothesis is rejected at Z_α level of significance. Similarly if

$$u - E(u) \leq Z_\alpha \times S.E.(u) \tag{4.3}$$

The deviation is not regarded significant at 5% level of significance. In other words, the deviation u – E(u) could have arisen due to fluctuations of sampling, and the data do not provide any evidence against the null hypothesis which may therefore be accepted at a level of significance.

If $|Z| \leq 1.96$, then the hypothesis H_0 is accepted at 5% level of significance. Thus the steps to be used in the normal test are as follows:

Table 4.3 The critical values (Z_α) for Z

Level of significance (%)	Two-tailed test	Right-tailed test	Left-tailed test
10	1.645	1.28	−1.28
5	1.96	1.645	−1.645
1	2.58	2.33	−2.33

1. Compute the test statistic Z under H_0.
2. If $|Z| > 3$, H_0 is always rejected.
3. If $|Z| < 3$, test its significance at certain level of significance. Table 4.3 provides some critical values of Z

Large Sample Test for Single Mean

A very important assumption underlying the tests of significance for variables is that the sample mean is asymptotically normally distributed even if the parent population from which the sample is drawn is not normal. If x_i ($i = 1, 2, \ldots, n$) is a random sample of size n from a normal population with mean μ and variance $\frac{\sigma^2}{n}$ the sample mean is distributed normally with mean μ and variance, i.e. $\bar{x} \sim N\left(\mu, \frac{\sigma^2}{n}\right)$. Thus to test the hypothesis, H_0: $\mu = \mu_0$ against the alternative H_1: $\mu \neq \mu_0$, the test statistic is as follows:

$$Z = \frac{\bar{x} - \mu_0}{\frac{\sigma}{\sqrt{N}}} \sim N, (0, 1) \qquad (4.4)$$

If σ^2 is unknown, then it is estimated by sample variance, i.e. $\sigma^2 = s^2$ (for large n).

Example 4.3 A sample of 900 members has a mean of 3.4 cm and standard deviation (SD) 2.61 cm of certain attribute. Is the sample drawn from a large population of mean 3.25 cm?

Solution *Here a large sample has been drawn from the population with mean 3.25 cm. We set the null and alternative hypothesis as* H_0: $\mu = 3.25$ *and* H_1: $\mu \neq 3.25$. *The sample mean,* $\bar{x} = 3.4$ *cm,* $n = 900$, $\mu = 3.25$ *cm and* $\sigma = 2.61$.

So under H_0,

$$Z = \frac{3.4 - 3.25}{\frac{2.61}{\sqrt{900}}} = 1.73$$

Since $|Z| < 1.96$, *we conclude that the data does not provide any evidence against the null hypothesis* H_0 *may therefore be accepted at 5% level of significance.*

Large Sample Test for Single Proportion
Example 4.4

In a sample of 1000 people, 540 are fish eaters and the rest are meat eaters. Can we assume that both fish and meat are equally popular at 1% level of significance?

Solution *It is given that* $n = 1000$ *and* $x =$ *the number of fish eaters* $= 540$.

$p =$ *sample proportion of fish eaters* $= \frac{540}{100} = 0.54$

$P =$ *Population proportion of fish eaters* $= \frac{1}{2} = 0.50$

H_0: *Both fish and meat are equally popular, i.e.* $P = 0.50$; H_1: $P \neq 0.50$

$$Z = \frac{p - P}{\sqrt{\frac{PQ}{n}}} = \frac{0.54 - 0.50}{\sqrt{\frac{0.5 \times 0.50}{1000}}} = 2.5321 \tag{4.5}$$

Since computed Z <2.58 at 1% level of significance therefore fails to reject H_0, *we conclude that fish and meat are equally popular.*

Small Sample Test of Significance for Single Mean

Suppose a random sample $x_1, \ldots, x_n$ of size ($n > 2$) has been drawn from a normal population whose variance s^2 is unknown. On the basis of this random sample, the aim is to test

$$\begin{aligned} H_0 : \mu &= \mu_0 \\ H_1 : \mu &\neq \mu_0(\text{two-tailed}) \\ \mu &> \mu_0(\text{right-tailed}) \\ \mu &< \mu_0(\text{left-tailed}) \end{aligned}$$

Test statistic $= \frac{\bar{x} - \mu_0}{\frac{s}{\sqrt{n}}} \sim t_{n-1}$, *where*

$$\bar{x} = \frac{1}{n}\sum\nolimits_{i=1}^{n} x_i,$$

$$s^2 = \frac{1}{n-1}\sum_{i=1}^{n}(x_i - \bar{x})^2$$

The computed value of t is compared with the tabulated value at 5 or 1% levels of significance and at ($n - 1$) degrees of freedom, and accordingly the null hypothesis is accepted or rejected.

Example 4.5 A farmer has stocked some quantity of seed of rohu (*Labeo rohita*) in a pond for its production. After 1 year he wants to know whether fishes have attained the weight of 800 g (as said by researched) or not? He wants to keep DO (dissolved oxygen) level more than 5 ppm during the period in his pond for proper growth. He collects water sample monthly to monitor the DO in the pond. After 1 year he caught

Table 4.4 DO value and fish weight of sampled fish from the pond

Sl no.	DO (ppm)	Weight (g)
1	4.94	820
2	7.09	634
3	5.74	781
4	4.99	739
5	5.78	619
6	4.17	837
7	7.01	963
8	4.12	923
9	6.01	771
10	3.66	891
11	5.08	771
12	6.64	970
13		868
14		913
15		788
16		830
17		726
18		705
19		990
20		717

20 fishes as sample using a gill net to measure the weight of fishes. Weight of fishes and value of DO are given in Table 4.4.

Solution Solution for DO: *Here the hypotheses to be tested are* H_0: *mean DO is equal to 5 ppm against* H_1: *mean DO is less than 5 ppm. Here,*

$\bar{x} = 5.44$; $n = 12$; $\mu_0 = 5$; $SD = 1.14$.

The value of the t-statistic is

$$t = \frac{544 - 5}{\frac{1.14}{\sqrt{12}}} = 1.32$$

Table value of t at 5% level and 11 degree of freedom is 2.2009. Since calculated t-value is less than tabulated t-value, hence based on sample we fail to reject H_0.

Therefore we can say that DO level in the pond during the year was almost 5 ppm or more.

Solution for weight: The hypotheses are H_0: the mean weight of captured fish is equal to 800 g, i.e. $\mu = 800$, and H_1: the mean weight of captured fish is not equal to 800 g, i.e. $H_0 \neq 800$. Here, sample mean $= \bar{x} = 812.8$; $n = 20$; $SD = 107.45$.

Under H_0,

$$t = \frac{8128.800}{\frac{107.451.14}{\sqrt{20}}} = 0.532$$

Table value of t at 5% level and 19 degree of freedom is 2.093. Since calculated t-value is less than tabulated t-value, hence based on sample we fail to reject H_0. Therefore, we can say that weight of rohu in the pond after 1 year was almost 800 g.

4.5.3 Parametric Test of Significance Based on Two Samples

Suppose we want to evaluate the effect of supplementary stocking on the fish yield of floodplain wetlands (beels) of Assam. This could be evaluated in two different ways: (1) In a planned experiment, we select a group of subjects (i.e. beels) and record their fish yield. Then they must be subjected to supplementary stocking through in situ raising of fingerlings in pens for a period, after which we record their yield again. We compare the yield before and after the adoption of technology. (2) In a sample survey, we consider two groups of beels: one group of beels not practising supplementary stocking regularly and another group of beels practising. We compare the yield of the beels in the two groups. These two approaches illustrate the two main techniques of comparing the two sets of data. This section delineates three parametric tests of significance based on two samples, that is, paired *t-test*, unpaired *t-test* for means and *F-test* for equality of variances.

Paired t-Test for Comparing Two Dependent Samples

The paired sample *t-test*, sometimes called the dependent sample *t-test*, is a statistical procedure used to determine whether the mean difference between two sets of observations is zero. In a paired sample *t-test*, each subject or entity is measured twice, resulting in pairs of observations. Common applications of the paired sample *t-test* include case-control studies or repeated-measures designs. Suppose you are interested in evaluating the effectiveness of fish stock enhancement through adoption of pen aquaculture. One approach you might consider would be to record the fish yield of a sample of beels before and after adoption of technology and analyse the differences using a paired sample *t-test*.

Assumptions for Paired *t-Test*

In a paired sample *t-test*, the observations are defined as the differences between two sets of values, and each assumption refers to these differences, not the original data values. The paired sample *t-test* has four main assumptions:

1. The dependent variable must be continuous.
2. The observations are independent of one another.
3. The dependent variable should be approximately normally distributed.
4. The dependent variable should not contain any outliers.

Hypotheses

H_0: The true mean difference (μ_d) is equal to zero, that is, $\mu_d = 0$.

H_1: The true mean difference (μ_d) is not equal to zero.

That is, $\mu_d \neq 0$ (two-tailed) or $\mu_d > 0$ (right-tailed) or $\mu_d < 0$ (left-tailed).

Test Statistic

Let n be the size of each of the two samples and d_i, $i = 1, 2, \ldots, n$ the difference between the corresponding members of the sample. Let $\overline{d}$ denote the mean of differences and the *SD* standard deviation of these differences. The test statistic is given by

$$t = \frac{\overline{d}\sqrt{n}}{S_d} \sim t_{n-1}$$

where, $\overline{d} = \frac{1}{n}\sum_{i=1}^{n} d_i$, $S_d = \sqrt{\frac{\Sigma\left(d_i - \overline{d}\right)^2}{n-1}}$

Test Criterion: Null hypothesis (H_0) is rejected at level α if $t_{calculated} > t_{(n-1)}$.

Example 4.6 A study was conducted to determine if there is a significant effect of adoption of fish stock enhancement through pen aquaculture technology on fish yield of beels of Assam. MS Excel snip shown below represents yield of 22 beels of Assam before and after the adoption of technology. The normality test for the differences was examined by *q-q* plot and Shapiro–Wilk test statistic ($W = 0.961$, $p = 0.501$). The *p-value* is not significant, and we conclude that differences (*di's*) follow normal distribution. Therefore, we can apply paired *t-test*.

Step 1: State the null and research hypotheses.

H_0: The mean yield of beels before and after the adoption of technology is the same, that is, $\mu_1 = \mu_2$.

H_1: The mean yield of beels before and after the adoption of technology is different, that is, $\mu_1 \neq \mu_2$.

Step 2: Compute the test statistic.

SPSS software has been used for analysing the above data. Feed the following data from Excel to SPSS. Go to Analyze → Compare Means → Paired Samples TTest.

Send variables after and before to the Variable 1 and Variable 2 under Paired Variables box by clicking the small arrow button. Finally, click OK. SPSS output obtained is shown in Table 4.5 and Figs. 4.3, 4.4, and 4.5:

Step 3: Statistical decision.

Table 4.5 Paired samples statistics

	Mean	*N*	SD
Before	570.3636	22	348.021
After	792.0454	22	394.0425

	A	B	C
1	Before (x_i)	After (y_i)	$d_i=y_i-x_i$
2	62	141	79
3	365	453	88
4	680	1007	327
5	324	333	9
6	327	836	509
7	648	767	119
8	457	983	526
9	398	773	375
10	389	613	223
11	578	492	-86
12	586	629	43
13	548	926	378
14	853	1307	454
15	335	485	149
16	479	748	269
17	840	857	17
18	1827	2102	275
19	486	646	161
20	671	661	-10
21	891	1157	265
22	234	717	483
23	570	792	222

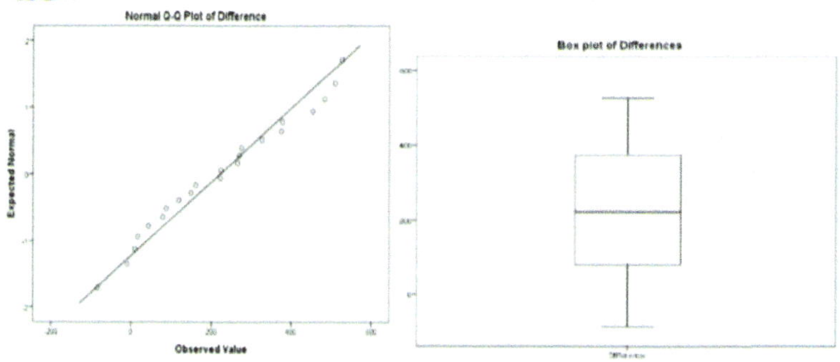

Fig. 4.3 Screenshot results of paired *t-test* in SPSS package

Here, the calculated *p-value* of the statistic is 0.001 which is lower than the observed *p-value* (0.05). Hence, it is highly significant, and we reject the null hypothesis and conclude that there is significant difference in the yield of beels.

Step 4: Interpret the result.

The paired *t-test* was highly significant ($t = 5.742$, $df = 21$, $p = 0.001$). We rejected the null hypothesis, suggesting that the fish mean yield of beels was significantly higher after (792.1 ± 394.1 kg/ha) intervention than that of before

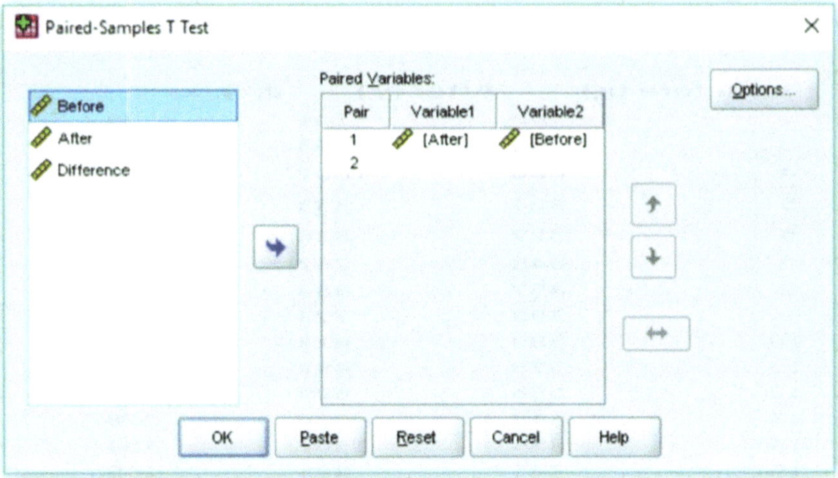

Fig. 4.4 Screenshot of paired *t-test* in SPSS package

	Paired Differences					t	df	Sig. (2taile d)
	Mean	SD	SE (Mean)	95% CI of the Difference				
				Lower	Upper			
Pairl After-Before	221.68	181.08	38.61	141.39	301.97	5.74	21	.000

Fig. 4.5 Screenshot of paired *t-test* in SPSS package

intervention (570.4 ± 348.1 kg/ha). The data shows that mean yield increased by 38.9% after intervention.

4.5.4 *t-Test* for Testing of Difference Between Two Means of Uncorrelated Observations

The independent sample student's *t*-test (also known as Welch two sample test) consists of comparing two mean values of continuous-level (interval/ratio data), normally distributed data. It requires that samples are independent, of equal variance and normally distributed. It tells us whether the difference we see between the two independent samples is a true difference or whether it is just a random effect caused by skewed sampling. The Shapiro–Wilk test may be employed to check for normality prior to performing the comparison; Fisher's *F-test* will help checking for equal variances and Levene's test for equality of error variances.

Let $\bar{x}$ and s_1 be the mean and standard deviation of a sample of size n_1 from a normal population with mean μ_1 and σ_1^2 variance, and let $\bar{y}$ and s_2 be the mean and standard deviation of another sample of size n_2 from a normal population with mean μ_2 and variance σ_2^2. We assume that $\sigma_1^2 = \sigma_2^2 = \sigma^2$. We want to test whether the population means differ significantly.

Hypothesis: H_0: The mean difference is zero, that is, $\mu_1 = \mu_2$. H_1: The mean difference is significantly different from zero, that is, $\mu_1 \neq \mu_2$ (two-tailed) or $\mu_1 > \mu_2$ (right tailed) or $\mu_1 \leq \mu_2$ (left-tailed).

Test statistic: Under H_0, the test statistic is given by

$$t = \frac{\bar{x} - \bar{y}}{S\sqrt{\frac{1}{n_1} + \frac{1}{n_2}}} \sim t_{n_1+n_2-2}$$

where $\bar{x} = \frac{1}{n}\sum_{i=1}^{n} x_i, \bar{y} = \frac{1}{n}\sum_{i=1}^{n} y_i$ and $S = \sqrt{\frac{\Sigma(x-\bar{x})^2-(y-\bar{y})^2}{n_1+n_2-2}}$

Test criterion: Null hypothesis (H_0) is rejected at the level α if $t_{\text{calculated}} > t_{n_1+n_2-2}$

.

Example 4.7 In a wetland (beel) survey, the fish production (MT) of randomly selected beels was recorded as under. 1 stands for unstocked beels and 2 stands for stocked beels. Test whether there is significant difference in mean production of stocked and unstocked beels. Shapiro–Wilk test statistic ($W = 0.895$, $p = 0.302$) conducted for test of normality showed that data is normally distributed.

Step 1: State the null and research hypotheses.

H_0: In the general population, the mean production of stocked and unstocked beels is the same.

H_1: In the general population, the mean production of stocked and unstocked beels is different.

Step 2: Compute the test statistic.

Feed the data into SPSS as shown in Excel snip above.

Go to Analyse → Compare Means → Independent Samples T Test. Send variables production to the test variable(s) box by clicking the small arrow button. Click grouping variable. New window define groups open up. Enter 1 for Group 1 and 2 for Group 2. Click continue to return to independent samples T Test... window. Finally, click OK. SPSS output obtained is shown in Fig. 4.6 and Tables 4.6, 4.7, and 4.8.

Step 3: Statistical decision.

Levene's test for equality of variances shows that error variances are not equal. Here, the calculated *p-value* of the statistic is 0.03 (for both equal and unequal variances assumed) which is low than observed *p*-value (0.05). Hence, it is significant, and we reject the null hypothesis and conclude that there is significant difference between production of beels.

Step 4: Interpret the result.

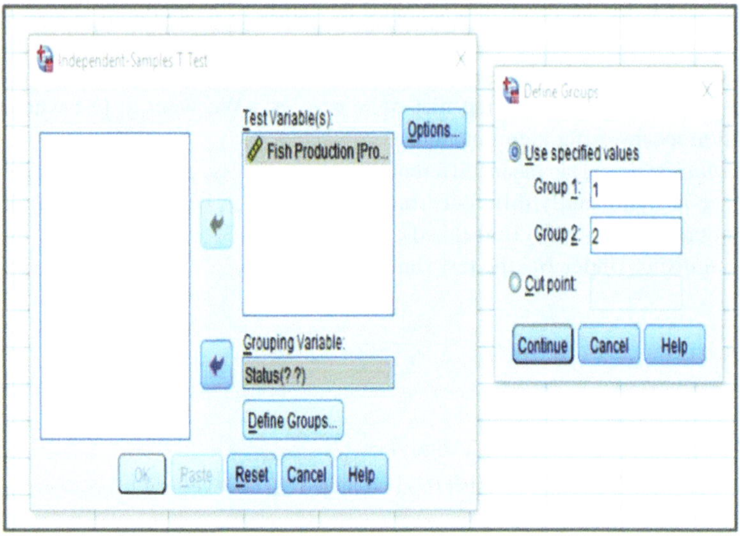

Fig. 4.6 Screenshot of two sample *t-test* in SPSS package

The independent sample *t-test* for unequal variances assumed was significant ($t = 2.55$, $df = 8$, $p = 0.03$). We rejected the null hypothesis, suggesting that the production of stocked beels (2120.03 ± 921.3 MT) was significantly higher than unstocked beels (1153.8 ± 391.6 MT).

4.6 F-Test for Equality of Two Population Variances

F-test was first worked out by G.W. Snedecor. This test is referred to as variance ratio test. F-test is used to test whether two independent samples (x_i, $i = 1, 2, \ldots, n_1$) and (y_j, $j = 1, 2, \ldots, n_2$) have been drawn from the normal distributions with the same variance σ^2. Alternatively, we can say that it is used to test whether the two independent estimates of the population variances are homogeneous. Let s_1^2 be the variance of a sample of size n_1 and s_2^2 be the variance of a sample of size n_2.

Hypothesis

H_0: The population variances are equal $\sigma_1^2 = \sigma_2^2 = \sigma^2$.

H_1: The population variances are not equal, that is, two-tailed or right-tailed or left-tailed.

Test Statistic

Under H_0, the test statistic is given by

Table 4.6 Screenshot of two sample *t-test* data in MS Excel

	A	B	C
1	Stocking_status	Production	
2	1	941.4	
3	1	616.3	
4	1	1048.3	
5	1	829.4	
6	1	1520.5	
7	1	1514.3	
8	1	1606.744	
9	2	1255.81	
10	2	1233.44	
11	2	1656.92	
12	2	1507.81	
13	2	2647.743	
14	2	3105.998	
15	2	3432.456	
16			

Table 4.7 Screenshot of SPSS output

Group statistics				
	Group	*N*	Mean	SD
Production	Stocked	7	1153.849	391.5723
	Stocked	7	2120.025	921.5275

Table 4.8 Screenshot of SPSS output

Independent samples test							
Levene's test for equality of variances	t-test for equality of means						
						95% CI of the difference	
F Sig.	*t*	df	Sig. (2-tailed)	Mean difference	Std. error difference	Lower	Upper
14.17	2.55	12.00	0.03	966.18	378.44	141.62	1790.74
0.00	2.55	8.10	0.03	966.18	378.44	95.32	1837.03

	A	B	C
1	n1=	30	
2	n2=	25	
3	s12=	60	
4	s22	40	
5	F (caculated)	1.5	=B3/B4
6	F(29. 24. 0.05)=	1.9453	=F.INV.RT(0.05,B1-1,B2-1)
7			

Fig. 4.7 Screenshot for F-test in excel

$$F = \frac{S_1^2}{S_2^2} \sim F_{(n_1-1),n_2-1)}$$

where $S_1^2 = \frac{\sum (x-\bar{x})^2}{(n_1-1)}$ and $S_2^2 = \frac{\sum (y-\bar{y})^2}{(n_2-1)}$

Note: The greater of the two variances is to be taken in the numerator.

Test Criterion

Null hypothesis is rejected at level α if $F_{calculated} > F_{(n_1-1),(n_2-1)}$.

Example 4.8 Data on daily fish landings recorded for 31 days in a landing centre showed variance of 60 kg^2, whereas landings recorded for 25 days in another centre showed variance of 40 kg^2. We want to test whether the variability of daily fish landings is the same in two landing centres.

Step 1: State the null and research hypotheses.

H_0: In the general population, there is no variability of daily fish landings in two landing centres.

H_1: In the general population, there is a variability of daily fish landings in two landing centres.

Step 2: Compute the test statistic.

The steps for computing the F-statistic is shown in the Excel snip given below (see Fig. 4.7).

Step 3. Statistical decision.

The critical (tabulated) value for rejecting the null hypothesis is 1.94, and the obtained (calculated) value is 1.5. Here if the critical value is not less than or equal to the obtained value, we may accept the null hypothesis. This implies that there is no variability of daily fish landings in two landing centres.

Step 4: Interpret the result.

The F-test was not significant ($F_{(19,24)} = 1.5, p > 0.05$). We may accept the null hypothesis and conclude that there is no variability of daily fish landings in two landing centres.

4.7 Conclusion

The researchers always make inferences on inland fisheries based on some representative sample data. This chapter provides a guideline on the basics of inferences by using a statistical test of significance. This chapter will help the reader to formulate the null and alternative hypothesis. The chapter guides the users to devise decision rule for accepting and rejecting the hypothesis along with the errors. The chapter describes only frequently used hypothesis testing. The example used will help the readers to devise their hypothesis formulation and their testing. The concept impregnated will help the readers in the prudent use of parametric tests, e.g. Z-test, t-test, paired t-test, χ^2 test, F-test, etc.

Exercise

1. A random sample of 30 fishes from a particular species showed an average length of 8.2 cm. Can you conclude that this has been drawn from a population with mean 8.1 cm and standard deviation 0.25 cm at 1% level of significance?
2. A total of 1250 fishes were taken at random from a pond, and their mean length was 9.95 cm with a standard deviation of 7.81 cm. Calculate an approximate 95% confidence interval for the mean length of the fishes.
3. In a particular water body, the long-term proportion of catla caught in the total catch is 51.46%. In a random sample of 5000 fishes caught from the same water body, the proportion of catla was 52.55%. Determine whether those two proportions differ significantly at 10% level or not.
4. The price of an aquarium fish feed at a national store is Rs. 179. A person has purchased the same feed at an online auction site for the following prices 155, 179, 175, 175 and 161 rupees. Determine (at 1% level of significance) whether the average price of the aquarium fish feed is than Rs. 179 if purchased at an online auction. Assume that the auction prices of fish feeds are normally distributed.
5. A group of 20 fishery graduate students were assessed by a test before and after a training programme. Their pre-test and post-test scores were recorded. The mean and standard deviation of the differences were 2.05 and 2.837, respectively. Assuming that the differences are normally distributed, can we conclude that the training improved the knowledge level of students.
6. A random sample of 15 fishes taken from a pond showed a mean length of 68.4 cm with a standard deviation of 16.47. A sample of 12 fishes taken from the same pond had a mean length of 83.42 cm and standard deviation of 17.63 cm. Can we conclude that there is no difference between the two means?
7. Failing to reject null hypothesis when it is false is
   (a) α
   (b) Type I error
   (c) Type II error
   (d) β

8. A 'statistic' is
   (a) A sample characteristic
   (b) A population characteristic
   (c) Unknown
   (d) Normally distributed
9. What is 'standard normal variate'?
10. What is the 'power' in hypothesis testing?

Further Reading

Anderson TW, Sclove SL (1974) Introductory statistical analysis. Houghton Mifflin Company, Boston

Bhujel RC (2008) Statistics for aquaculture, 1st edn. Wiley, Hoboken

Biradar RS (2002) Fisheries statistics, 2nd edn. Central Institute of Fisheries Education, Mumbai

Clarke GM, Cooke D (1998) A basic course in statistics. Arnold. Electronic Statistics Textbook. http://www.statsoftinc.com/textbook/stathome.html

Sheskin D (1997) Handbook of parametric and nonparametric statistical procedures. CRC Press, Boca Raton, p 719

Freund JE (2001) Modern elementary statistics. Prentice-Hall, Hoboken

Gupta SC, Kapoor VK (2006) Fundamentals of mathematical statistics. New Delhi, Sultan Chand & Sons, Educational Publishers

Johnson RA, Bhattacharyya GK (1992) Statistics: principles and methods, 2nd edn. Wiley, Hoboken

McDonald JH (2014) Handbook of biological statistics, 3rd edn. Sparky House Publishing, Baltimore

Mood AM, Graybill FA, Boes DC Introduction to the theory of statistics. Tata McGraw-Hill, New Delhi

Rohatgi VK, Saleh AK An introduction to probability and statistics. Wiley, New York

Roy AK, Sarangi N (2008) Applied bioinformatics, statistics & economics in fisheries research. New India Publishing Agency, New Delhi

Chi-Square Test of Significance

5

Key Topics
- Concept of goodness–of–fit and Chi–square test
- Chi–square test for the strength of association
- Chi–square test for independence

Test of significance provides an objective procedure for distinguishing between whether the observed difference signifies any real difference among groups. It indicates whether observed differences between assessment results occur because of sampling error or chance. The experiments in fisheries science are affected by a substantial amount of uncontrolled variations making such tests necessary. Sometimes data are best collected or conveyed nominally or categorically. These data are represented by counting the number of times a particular event or condition occurs. There are many instances in inland fisheries research, wherein nominal/categorical data describe the phenomenon under investigations more adequately than interval/ ratio data. Chi-square, a non-parametric test of significance, is an appropriate test when the data are in the form of frequency counts occurring in two or more mutually exclusive categories (nominal variables). It enables us to decide on the basis of sample if (1) a given set of counts (or frequencies) statistically match some known, or expected, set or (2) two or more categories are statistically independent. In this article, test of significance based on chi-square is presented with the examples on inland fisheries data. The data used in the article are analysed using MS Excel/SPSS.

Assumptions for Chi-Square
1. Each observation must fall in one and only one category.
2. The observations in the sample are independent of one another.
3. The observations are measured as frequencies.

B. K. Das et al., *Concept Building in Fisheries Data Analysis*,
https://doi.org/10.1007/978-981-19-4411-6_5

4. The expected frequency for each category must not be less than 5 for $df \geq 2$ and not less than 10 for $df = 1$.
5. The observed χ^2 with one degree of freedom must be corrected for continuity in order to use the table of values of critical.

5.1 Chi-Square Test for Goodness of Fit (One Sample Chi-Square)

Some situations in fisheries research involve investigations and questions about relative frequencies and proportions for a distribution, for example, a comparison of the number of women fishers to the number of men fishers involved in fishery-related activities, or a search for significant changes in the proportion of floodplain wetlands (beels) adopting supplementary stocking over 5 years, or an analysis of consumers' preference for different fish species in a state. In each of these examples, a question about a proportion in the population arises. The chi-square goodness-of-fit test uses sample data to test the hypothesis about the proportions of the population distribution. It is primarily used for testing the discrepancy between the expected and the observed frequency.

Hypotheses

H_0: The observed distribution of frequencies equals the expected distribution of frequencies in each category.

H_1: The observed distribution of frequencies does not equal the expected distribution of frequencies.

Test Statistic If O_i *and* E_i, $i = 1, 2, \ldots, n$ refers to observed frequency (the date) and expected frequency (the hypothesis), then χ^2 statistic is given by

$$\chi^2 = (O_i - E_i)^2 / E_i \sim \chi^2_{(n-1)}$$

The expected frequency is determined by $E_i = P_i n$, where P_i is a category's frequency proportion with respect to the other categories and $\Sigma O = n$ is the sample size of all categories.

Test Criterion Null hypothesis (H_0) is rejected at level α if $\chi^2_{\text{calculated}} > \chi^2_{\alpha,(n-1)}$.

Example 5.1

A survey was conducted to determine if there is a significant preference for a type of fish species purchased by consumers at selected fish markets of Tripura. Table 5.1 represents observed frequencies for the 160 participants who were surveyed at selected fish markets of Tripura (Apu Das et al. 2013). We want to determine if there is any preference for purchasing one of the six fish species by the consumers of Tripura.

Table 5.1 Observed frequencies for different fish species in Tripura

Species	Frequency
Rohu	67
Catia	42
Carpio	16
Hilsa	13
Migaia	9
Others	13

Step 1: State the Null and Research Hypotheses

$H_{0:}$ In the population of fish consumers of Tripura, there is no preference of one fish species over any other.

$H_{0:}$ In the population of fish consumers of Tripura, there is at least one fish species preferred over the others.

Step 2: Compute the Test Statistic

Under null hypothesis, the expected frequency, E_i, will be the same for all six categories, since our research problem assumes that all categories are equal.

That is,

$$E_i = P_i n = \frac{1}{6}(160) = 26.7$$

	A	B	C	D	E	F
1						
2	**Species**	**O_i**	**E_i**	**$(O_i\text{-}E_i)^2/E_i$**		
3	Rohu	67	26.7	60.82734	**=((B3-C3)^2)/C3**	
4	Catla	42	26.7	8.76742		
5	Carpio	16	26.7	4.28801		
6	Hilsa	13	26.7	7.02959		
7	Mrigal	9	26.7	11.73371		
8	Other fishes	13	26.7	7.02959		
9		Chisquare calculated=		99.67566	=SUM(D3:D8)	
10		Chisquare tabulated=		11.07049769	=CHISQ.INV.RT(0.05,5)	
11			p-value=	6.18605E-20	=CHISQ.DIST.RT(D9,5)	
12						
13						

Step 3: Statistical Decision

The critical (tabulated) value for rejecting the null hypothesis is 11.07, and the obtained (calculated) value is 99.68. If the critical value is less than or equal to the obtained value, we must reject the null hypothesis.

Step 4: Interpret the Result

The chi-square goodness-of-fit test was highly significant $\chi^2_5 = 99.67,\ p < 0.001$). We rejected the null hypothesis, suggesting that there is a real difference among fish species preferred by consumers of Tripura. The data shows that a larger portion of the consumers preferred the rohu fish, and only a few of them preferred the mrigal.

5.2 Chi-Square Test for Independence

Some research involves investigations of frequencies of statistical associations of two categorical attributes. The chi-square statistic can be used when two or more categories are involved for two attributes. Suppose that an attribute data of size N is classified according to two attributes, say A and B, and the attribute A is further subdivided into r classes $A_1, A_2, \ldots, A_r$ and the attribute B into $B_1, B_2, \ldots, B_c$. Such attribute data can be presented in the form of a table called r x c contingency table as shown below:

Attribute	B_1	B_2	...	B_j	...	B_c	Total
A_1	O_{11}	O_{12}	...	O_{1j}	...	O_{1c}	$A_{1.}$
A_2	O_{21}	O_{21}	...	O_{2j}	...	O_{2c}	$A_{2.}$
...	...	...	...	...	...	...	...
A_i	O_{i1}	O_{i2}	...	O_{ij}	...	O_{ic}	$A_{i.}$
...	...	...	...	...	...	...	...
A_r	O_{r1}	O_{r2}	...	O_{rj}	...	O_{rc}	$A_{r.}$
Total	$B_{.1}$	$B_{.2}$	...	$B_{.j}$	...	$B_{.c}$	$A_{..} = B_{..} = N$

Hypotheses: Statement of Hypothesis

$H_{0:}$ The observed distribution of frequencies equals the expected distribution of frequencies in each category.

$H_{1:}$ The observed distribution of frequencies does not equal the expected distribution of frequencies.

Test Statistic If O_{ij} and, $E_{ij} = 1, 2, \ldots, r; j = 1, 2, \ldots,$ c refers to observed frequency (the data) and refers to the expected frequency (the hypothesis) corresponding to ith row and jth column of the $r \times c$ contingency table, then χ^2 statistic is given by

$$\chi^2 = \sum_{j=0}^{c} \sum_{i=1}^{r} \frac{\left(O_{ij} - E_{ij}\right)^2}{E_{ij}} \sim \chi^2_{(r-1)(c-1)}$$

The expected frequency is determined by $E_{ij} = \frac{A_{i.} \times B_{.j}}{N}$, where $A_{i.}$ and $B_{.j}$ are total frequency of class corresponding to ith row and jth column and N is the total of all categories and $\Sigma\Sigma O = N$.

Test Criterion Null hypothesis (H_0) is rejected at level α if.

A simple formula for computing using the Yates correction of a 2 × 2 *contingency table is given by.*

Attribute	A_1	A_2	Total
B_1	a	b	$a + b$
B_2	c	d	$c + d$
Total	$a + c$	$b + d$	$a + b + c + d = n$

Example 5.2

In a series of experiments to test whether advanced stages of *Myxobolus* infection is cured by lime treatment, the observations found are given in the table below. We want to test whether lime has any effect in curing the infection.

	Untreated	Treated	Total
Cured	55 (a)	77 (b)	132 ($a + b$)
Uncured	47 (c)	33 (d)	80 ($c + d$)
Total	102 ($a + c$)	110 ($b + d$)	212 (n)

Step 1: State the Null and Research Hypotheses

H_0: In the general population, there is no association between the lime treatment and curing of *Myxobolus* infection.

H_1: In the general population, there is a predictable relationship between the lime treatment and curing of *Myxobolus* infection.

Step 2: Compute the Test Statistic

The expected frequency is calculated as follows. The steps for computing the χ^2 statistic with Yates correction are shown in the Excel snip given below.

	A	B	C	D	E
1		Untreated	Treated	total	
2	Cured	55	77	132	
3	Unncured	47	33	80	
4		102	110	212	
5					
6	Chi-square statistic with Yates correction=			5.15887	=(D4*(ABS(B2*C3-C2*B3)-(D4/2))^2)/(D2*D3*B4*C4)
7			Chi-square tabulated=	3.841459	=CHINV(0.05,1)
8			p-value=	0.023128	
9					

Step 3: Statistical Decision

The critical (tabulated) value for rejecting the null hypothesis is 3.841, and the obtained (calculated) value is 5.159. If the critical value is less than or equal to the obtained value, we must reject the null hypothesis. The rejection of null hypothesis implies that *Myxobolus* infection is affected by the lime treatment.

Step 4: Interpret the Result

The chi-square goodness-of-fit test was significant ($\chi_1^2 = 5.159,\ p < 0.05$). We rejected the null hypothesis, suggesting that infection is cured by lime treatment. The data shows that lime increased the curing of *Myxobolus* infection from 53.9 % ($55 \div 102 \times 100$) to 70.0 % ($77 \div 110 \times 100$), that is, 16.1% improvement.

Example 5.3

In a fish tagging experiment, the length frequency of tagged fishes and recoveries was as under. Test whether the length distributions can be accepted as the same.

Length(cm)	Fishes tagged	Fishes recaptured
10–20	108	9
20–30	140	15
30–40	256	28
40–50	358	40
50–60	111	8
Total	1000	100

Step 1: State the Null and Research Hypotheses

$H_{0:}$ In the general population, length distribution of tagged and recovered fishes is the same.

$H_{1:}$ In the general population, length distribution of tagged and recovered fishes is not same.

Step 2: Compute the Test Statistic

We shall use the SPSS software to analyse the above data. Feed the following data from Excel to SPSS. The chi-square test for independence requires one variable to identify the conditions of the rows and a second variable to identify the conditions of the column. Here 'Tagging' variable represents the rows, and 'Group' represents the column. 'Frequency' represents the observed frequencies. Assign values to serve as a reference for the column and row variables. The Variable Tagging is labelled as 1 = Fishes tagged and 2 = Fishes recovered, and variable Length Group (cm) is labelled as 1 = 10–20, 2 = 20–30, 3 = 30–40, 4 = 40–50, and 5 = 50–60.

Go to Data ⇒ Weight Cases ... The default setting is **Do not weight cases**. Click the circle next to **Weight cases** as shown in the figures on next page Move that variable Frequency to the **Frequency Variable** box by clicking the small arrow button. Finally, click **OK.**

	A	B	C	D
1	Tagging	Group	Frequency	
2	1	1	108	
3	1	2	140	
4	1	3	256	
5	1	4	385	
6	1	5	111	
7	2	1	9	
8	2	2	15	
9	2	3	28	
10	2	4	40	
11	2	5	8	
12				

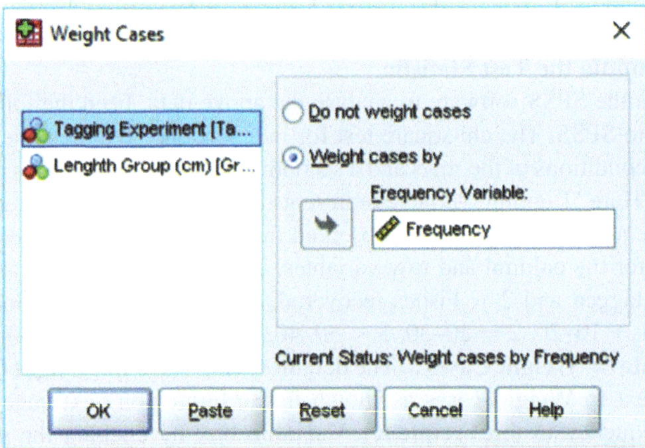

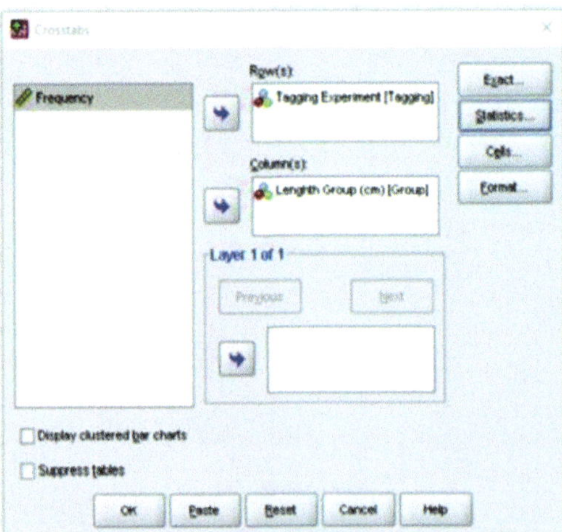

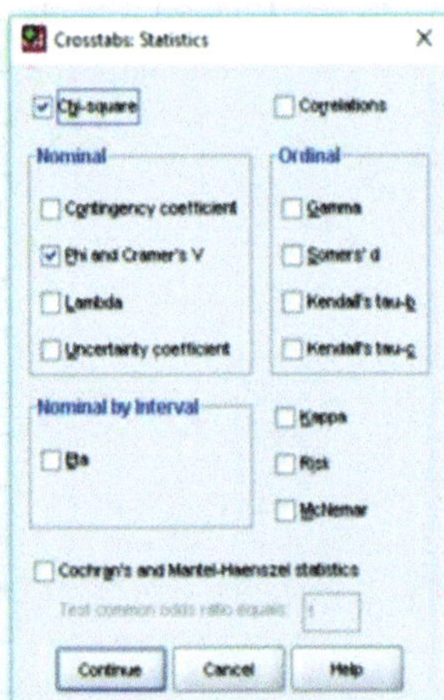

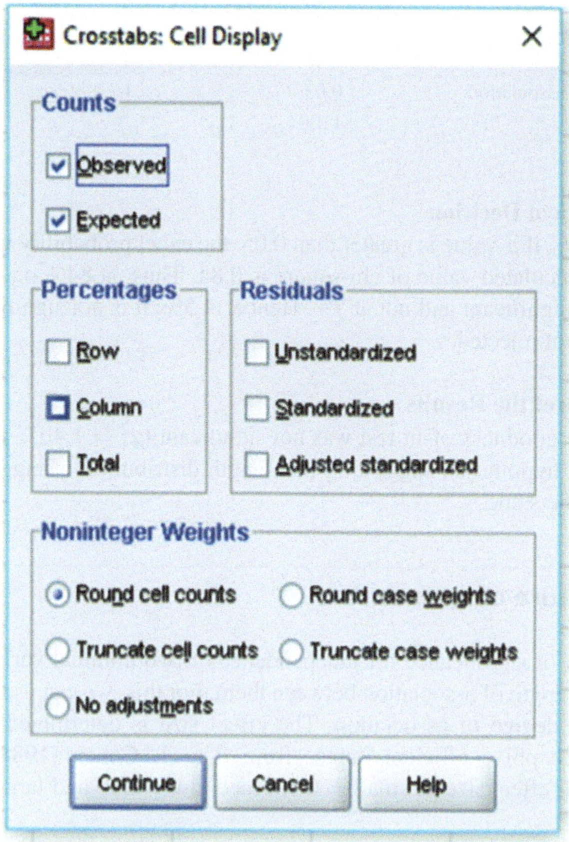

For analysis of data, Go to the menu.

Analyze ⇒ Descriptive Statistics ⇒ Crosstabs...
In Crosstabs window, move Tagging Experiment variable under **Row(s)** box and Length Group (cm) under **Column(s)** box. Click **Statistics**χ in Crosstabs window; the new window Crosstabs: Statistics opens. Check the box **Chi-square**, and then click **Continue** to the return Crosstabs window. Again click **Cells**... in Crosstabs window, and the new window Crosstabs: Cell Display opens. Check the boxes **Observed** and **Expected** under **Counts.** Then, click **Continue** to return to the Crosstabs window. Finally, click **OK** to perform the analysis. SPSS output obtained is presented in Table 5.2.

Table 5.2 Output of chi-square analysis in SPSS

	Value df approx. sig. (two-sided)		
Pearson chi-square	1.40	4	0.843
Likelihood ratio	1.48	4	0.829
Linear-by-linear association	0.03	1	0.862
N of valid cases	1100		

Step 3: Statistical Decision

Do not reject H_0, if p-value is greater than 0.05, the exact probability value (p-value) at which the calculated value of chi-square is 0.84. Thus, at 84% onwards only the test statistic is significant and not at 5%. Hence, at 5%, it is not significant. The null hypothesis is not rejected.

Step 4: Interpret the Results

The chi-square goodness-of-fit test was not significant ($\chi^2_4 = 1.405,\ p > 0.05$). We accept the null hypothesis, suggesting that length distribution of tagged and recovered fishes is the same.

5.3 Measure of Association

Chi-square test of significance for independence of two nominal variables does not describe the strength of association between them. For this, we can use the effect size to analyse the degree of association. The effect size is determined using phi (ε) coefficient. The phi coefficient ranges from 0 to 1. Cohen (1988) defined the conventions for effect size as small = 0.10, medium = 0.30 and large = 0.50.

Contingency table	Statistic for measure of association
2 × 2	phi (φ) = $\sqrt{\frac{\chi^2}{n}}$
>(2 × 2)	Cramer's, $V = \sqrt{\frac{\chi^2}{n(L-1)}}$

Note *χ^2 is the chi-square test statistic, n is the number in the entire sample, and L is the minimum value of row and column total from the contingency table.*

5.4 Conclusion

Some of the tests of significance based on chi-square and their application in inland fisheries were discussed in this chapter. The main focus was on test of significance using chi-square for nominal variables and measure of association between nominal variables. The procedure of carrying out analysis based on chi-square using MS Excel and SPSS was also discussed. This will help in analysing the measure of association between categorical data. For further study, one can access the references given below.

Exercise

1. A haul of fish contains 80 fishes comprising 4 different fish species. If all four species of fishes were equally likely to be caught in the haul, what would be the expected number of fishes of each species?
   (a) 20
   (b) 19
   (c) 18
   (d) 9
2. .... data is required for chi-square test?
   (a) Nominal
   (b) Categorical
   (c) Interval
   (d) Ratio
3. Goodness of fit of a distribution is tested by
   (a) *t*-test
   (b) Chi-square test
   (c) *F*-test
   (d) Sign test
4. What two elements are required to obtain the chi-square critical value from a table?
   (a) α, df
   (b) α, *p*-value
   (c) *p*-value, df
   (d) None of the above
5. For a $r \times c$ contingency table, the degree of freedom for the chi-square test statistic is
   (a) $r \times c$
   (b) $(r - 1) \times (c - 1)$
   (c) $n - 1$
   (d) $(r - 1) + (c - 1)$
6. If the calculated value of chi-square test statistic is less than its degrees of freedom, then which of the following is true for null hypothesis
   (a) Accept
   (b) Reject
   (c) No decision
   (d) None of these
7. In a chi-square test, the expected frequencies are computed under
   (a) Null hypothesis
   (b) Alternative hypothesis

(c) Both a and b
(d) None of these

8. For dichotomized frequencies, the degree of freedom for chi-square test statistic under null hypothesis is
(a) 3
(b) 2
(c) 0
(d) 1
9. If expected frequencies are less than 5, the test to be used is
(a) Pearson correlation
(b) Likelihood ratio
(c) Fisher's exact test
(d) None of these
10. A fisherman caught 8 catla, 5 rohu, 12 mrigal and 15 gonius using cast net in a pond. If all the species were equally likely to be caught in the net, what would be the value of chi-square test statistic?
(a) 5.4
(b) 10.0
(c) 6.2
(d) 5.8
11. An organization has the percentages of landings of major fish groups (IMCs, minor carps, catfishes, exotic carps, featherbacks) of a river for a decade ago. After noticing a drastic decline in IMCs landings, the organization carried out ranching of seed of IMCs in different stretches of River Ganga to rejuvenate the depleting stock. What statistical test is suitable to take decision?
(a) Chi-square test of independence
(b) *t*-test
(c) Chi-square of goodness of fit
(d) None of the above
12. A study was conducted on a sample of 200 fish to determine the sex ratio. Observed male and female were 130 and 70, respectively. Do the observed data fit the expected ration of 1:1?
(a) True at 5% level
(b) False at 5% level
13. A study was conducted to examine the impact of application of lime upon subsequent parasitic infection. The following table depicts the data.

	Parasite infection	
Lime	Yes	No
Yes	18	658
No	28	656

Which of the following statement is true?

(a) Parasite infection is associated with the application of lime.
(b) There is no association between parasite infection and lime application. (C) None of the above.

14. Five fishing gears were placed in randomly selected positions in a floodplain wetland (beel). The number of fishes captured in each gear in 1 day is given below.

Gear	A	B	C	D	E
No. of fishes	20	28	19	8	23

Is it statistically significant that the fish capture depends on fishing gears in this setting?
(a) Not statistically significant at 5%level of significance
(b) Statistically significant at 5% level of significance

15. The flow rate of a river is compared to the type of river bed at 50 different sites to investigate if there is an association between them, and the following table shows the relevant data. Which is the correct statement?

	High vegetation	Some vegetation	Sand	silt
Sloe flow	10	8	2	7
High flow	2	6	10	5

(a) The association cannot be examined statistically
(b) There is no statistically significant association at 5% level
(c) There is statistically significant association at 5% level
(d) None of the above

16. A study was conducted in randomly selected 40 floodplain wetlands (beels) to compare the type of connectivity (three categories seasonally open/open/closed) with stocking status of floodplain wetlands (two categories stocked/unstocked).

	Open	Seasonally open	Closed
Stocked	2	8	10
Unstocked	9	7	4

Perform a chi-square test to examine whether stocking status is independent of wetland connectivity or not.

17. The data below shows the results of a survey in which 250 fishers were classified according to level of education and their socio-economic status.

	Socio-economic status		
Education	Low	Medium	High
Primary	40	25	5
Secondary	40	20	5
Graduate	30	15	30
Post-graduate	15	15	10

Test whether the two criteria of classification are independent or not.

Further Reading

Freund JE (2001) Modern elementary statistics. Prentice-Hall
Bhujel RC (2008) Statistics for aquaculture, 1st edn. Wiley
Biradar RS (2002) Fisheries statistics, 2nd edn. Central Institute of Fisheries Education, Mumbai
Clarke GM, Cooke D (1998) A basic course in statistics. Arnold
Corder GW, Foreman DI (2009) Non-parametric statistics for nonstatisticians: a step-by-step approach. Wiley
Johnson RA, Bhattacharyya GK (1992) Statistics: principles and methods, 2nd edn. Wiley

Use of Non-parametric Test in Fisheries Data

6

Key Topics

- Concept of non-parametric test
- Data manipulation
- Hypothesis formulation
- Single sample problem
- Paired sample comparison
- Two independent samples comparison
- Three or more samples comparison

Descriptive statistics usually relates only to summarize or characterize the data and presentation of figures (visual or conceptual). The **descriptive statistics** may be a mean, median, mode, variance, range, histogram, etc., where no assumption(s) is made. Naturally, interpretations are mainly restricted to the observed samples. Thus descriptive statistics provide some information on the population. When the sample observations are utilized to make some inference on the population, then the subject area falls under **statistical inference** that constitutes two subject areas: estimation and testing of hypothesis. In this chapter, we focus only on the testing of hypotheses.

6.1 Concept of Non-parametric Test

The classical techniques of testing of hypotheses, statistical inference in broader sense, are based on specific assumption regarding the nature of the underlying population distribution. Usually its form and some parameter must be stated. A **parameter** is generally defined to characterize the probability distribution of the population. Elegant test statistics can be developed when the right kind of assumptions is made. In most of the cases, normal distribution is assumed to

B. K. Das et al., *Concept Building in Fisheries Data Analysis*,
https://doi.org/10.1007/978-981-19-4411-6_6

characterize the population, where two parameters are involved. Thereafter, complicated mathematics is applied to derive the properties of test statistics, which ultimately facilitate to carried out in the aforementioned approach is called **parametric test**. When it is used to make inference on the parameters (estimation or testing of hypothesis), then we call this method as parametric inference. These techniques are, however, exactly valid only so long as the assumptions themselves can be substantiated. The inference conclusions stated are valid only when the assumptions hold true, e.g. 'If the population is normal, then . . . .'. Otherwise, the inference may lead to wrong data interpretation.

Typically, **parametric** inferences require samples that are randomly drawn from a normally distributed population; consist of independent observations, except paired values; have respectiwve populations of approximately equal variances; are adequately large; and approximately resemble a normal distribution.

If any randomly drawn samples from population break one of those rules, assumption(s) of parametric test are violated.

However, in real-life situations, aforementioned assumptions may not hold true. In particular, violations of those assumptions are very often encountered in inland fisheries data. There are many such examples in the inland fisheries where distribution of a characteristic is not normal or standardized length distribution of fish captured by certain gear is far from normal and observed to be highly skewed. Species abundance in fish biodiversity assessment is generally positively skewed. Fish yield of wetlands is highly sparse and does not follow standard probability distributions. Moreover, inland fisheries are assessed on the basis of data emanated from exploratory survey in which outlying observations are very often encountered. Sometimes, few observations are recorded due to huge cost and manpower involved in survey on parasite infection or disease infection in fishes. In such cases, distribution of incidence of diseases or occurrence of parasites is clearly not normal. There are some basic options, as stated below, to handle these cases for applying parametric inferences.

- If there are outliers, it is likely that those have not come from the same population as the others. If the sample size is large, the easy way is to discard those observations and apply parametric test. If those outliers are in a small data set, alternative approach may be adopted.
- In many cases, data transformation, such as logarithmic, box-cox, square root, etc., is made to fulfil the assumptions. But data interpretations in terms of original parameters become difficult. In these situations, parametric inference may be inaccurate.
- In some settings, other standard distributions replace the normal distributions as models for overall pattern in the population. There are parametric inference methods for non-normal models.
- One may change the nature of the study to adhere to the rules. For example, one may use ordinal or nominal measurement scale or interval ratio scale. One may try to seek additional observation to enlarge the sample size. One may consult statistician for more advanced methods of parametric inferences.

Fortunately, alternative set of techniques is available. They are classified as distribution-free and non-parametric procedures. In non-parametric inference, whether estimation or testing of hypotheses, the methods are based on functions of sample observations. For example, actual observations are replaced by their relative ranks that are used to derive the test statistic for hypothesis testing. The specific distributional assumptions regarding the underlying population are not necessary. Instead non-parametric test are carried out under very general assumptions. The assumption most frequently required is that the population be continuous. More restrictive assumptions, like population is symmetric, are sometimes used. Unlike parametric test, non-parametric test implies a test for a hypothesis, which is not a statement about parameter values. Thus non-parametric test provides techniques for solving new kind of problems. On the other hand, non-parametric tests relate to an alternative approach to solve standard statistical problems, some of which are delineated in this chapter. At this juncture, it is pertinent to mention the advantages and disadvantages of non-parametric test as follows:

6.1.1 Advantages of Non-parametric Hypothesis Test

Non-parametric methods may be used on a greater variety of data because they require fewer conditions than their parametric counterparts. For this reason, it is less likely that non-parametric hypothesis tests will be performed inappropriately.

- Non-parametric procedures minimize the opportunities of inappropriate and indiscriminate applications, because the assumptions are so general that they are easily satisfied.
- Non-parametric methods are quick and easy to apply, since they involve very simple arithmetic that can be computed manually. For certain non-parametric procedures, the manual computations tend to be easier than their parametric counterparts.
- The derivations of non-parametric tests are easier to understand than their parametric counterpart, especially for non-mathematically trained users of statistics. Thus this lessons the danger of misuse of parametric statistical tools rely upon stringent assumptions.
- In the case of sparse data, which are very common in inland fisheries, most non-parametric techniques are relatively much more appropriate than parametric methods.
- Since most of the techniques are based on ranks, the scope of application is wider. For example, when recording of quantitatively data is not possible, the ranks will serve the purpose.

6.1.2 Disadvantages of Non-parametric Hypothesis Test

- Non-parametric hypothesis tests are less efficient than parametric tests. This means that, for a given level of significance, non-parametric tests require a larger sample size to reject a null hypothesis.
- Non-parametric tests replace the actual data values with either signs (positive or negative) or ranks. Thus, the exact data values are wasted. For example, non-parametric sign test is performed by replacing actual data values with positive or negative signs only.
- Because the use of non-parametric hypothesis tests is less widespread, graphing calculators and statistical software often do not have dedicated procedures for these tests.
- Use of non-parametric is less powerful and robust when the distributional assumption of the population is fulfilled.

This chapter describes commonly used non-parametric tests which are illustrated by the examples of inland fisheries data. Before embarking on systematic procedures to apply non-parametric test, one must know some basic data exploration and manipulation tools. These are described in the following section. The non-parametric test procedures and application of those procedures with fisheries data examples will be demonstrated in the subsequent sections.

6.2 Data Diagnostics for Using Non-parametric Test

One of the important assumptions is that sample is drawn randomly from normally distributed population. Data gathering process will ensure the random sample. However, examining normality of data samples requires some data exploration techniques, when size of the sample is fairly large. A graphical representation of data, such as box-whisker plot and histogram plot, is the simplest tool for visual inspection of normality. However, quantitative measures of skewness and kurtosis can also be used to detect normal distribution of data. There are several statistical tests (e.g. Shapiro–Wilk test) to examine the normality of data samples. In Example 6.1, data sample's diagnostics for normal distribution has been examined by using graphical methods.

Example 6.1
A survey was conducted in two districts of a state. Data on several features of wetlands was gathered. For demonstration purpose, we have considered only the yields of wetlands as continuous measured variables. One of the objectives was to compare the fish yield between two districts. In Fig. 6.1, histogram overlaid with normal distribution of fish yield of two districts has been presented. It clearly indicates violation of assumption of normal distribution of fish yield.

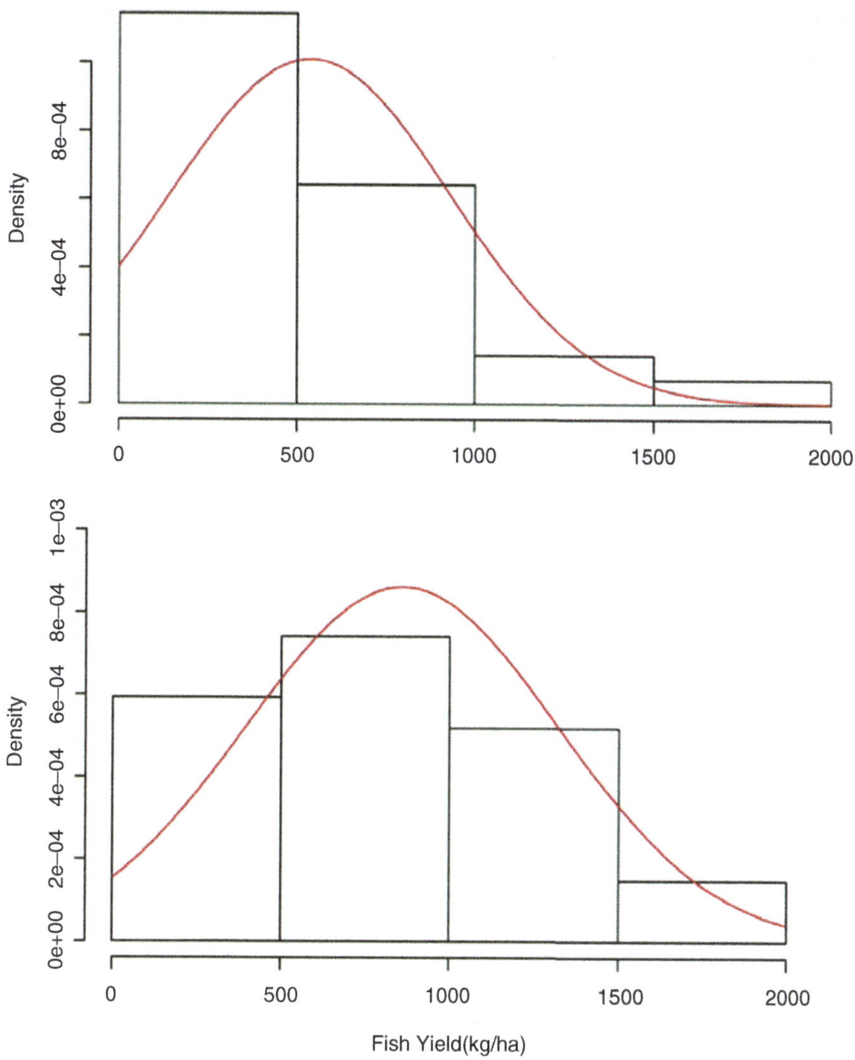

Fig. 6.1 Fish yield distribution of two districts

6.3 Ranking of Data

The most of the non-parametric test procedures involve derivation of test statistics that are function of data samples. Specifically, widely applicable non-parametric tests are.

Table 6.1 Fish species richness in river 1 and river 2

Number of fish species in river 1			93	83	79		73	
Number of fish species in river 2			87	96	92		84	
Ranks after combining fish species richness of two rivers								
Species richness	73	79	83	84	87	92	93	96
Rank	1	2	3	4	5	6	7	8

Table 6.2 Fish species richness in river 3 and river 4

River 3	90	85	95	70
River 4	75	80	55	90

constructed on the basis of relative rank data values. It is pertinent to learn how to convert the data sample values to rank values according to magnitude of data values. Ranking of data values is very simple and is demonstrated by the following example.

Example 6.2
Suppose a researcher wants to find out whether the fish species richness, which is measured as 'the number of species', in river 1 is higher than in river 2. The number of fish species recorded in river 1 and river 2 at 4 different sites in each of the river is shown in Table 6.1.

To rank all the values from Table 6.1 together, place them all in order in a new table from smallest to largest. Assign rank 1 to the first value; assign rank 2 to the second value; and so on. Note that all the values for river 2 are marked with red colour. Apparently it appears that the number species in river 2 is higher than that in river 2. ***However, if we want to seek for statistical significance, some type of procedure is needed. Those procedures will be described later on this chapter.***

6.3.1 Ranking with Ties

The aforementioned ranking procedure is simple and straightforward. In many cases, however, two or more of the data values are the same. In biodiversity assessment, for instance, this happens very frequently for 'species richness', a measure biodiversity. This implies that some values are repeated. We call this as the case of repeated values ties or tied values. Ranking of data values with ties needs separate treatment. Let us assume that fish species richness (the number of distinct fish species) recorded in river 3 and river 4 is shown in Table 6.2. Rank the values as in the previous example.

Note that the value 90 has been repeated. This means that the value of 90 is a tie. In the case of tie, assign all the tied values by the average of their rank values (Table 6.3). If these two values were different, they would be ranked 6 and 7. In this example, the average of 6 and 7 is 6.5, as shown in Table 6.4. Most of the non-parametric tests required different formula for the case of ties. Algebra behind the derivation of formula with ties is complex. Different test statistic is used for tied cases, though it produces slightly different results from the case of no ties.

Table 6.3 Ranking fish species richness with ties

Number of species	55	70	75	80	85	90	90	95
Rank ignoring ties	1	2	3	4	5	6	7	8
Rank with tie break	1	2	3	4	5	6.5	6.5	8

Table 6.4 Most popular non-parametric tests and their parametric equivalent for different analyses

Type of analysis	Non-parametric test	Parametric equivalent
One sample problem for location test	Wilcoxon signed rank test	*z*-test or *t*-test
Comparing two related	Wilcoxon signed rank	Paired *t*-test for dependent
Samples	Test	Samples
Comparing two unrelated samples	Mann-Whitney *U*-test, (Wilcoxon ranked sum test)	*t*-test for independent samples
Comparing three or more unrelated samples	Kruskal–Wallis test	One-way analysis of variance (ANOVA)
Comparing categorical data	Chi-square test	None
Examining a sample for randomness	Runs test	None

However, most of the modern software handle the tied cases automatically.

6.4 A Systematic Guide to Apply Non-parametric Test Procedures

A step-by-step general framework to carry out non-parametric test has been presented in this section. Before proceeding the details of individual test procedure, the list of frequently used non-parametric tests and their equivalent parametric counterparts is presented in Table 6.4. When demonstrating each of the non-parametric procedure, we will use a particular step-by-step method. A general step-by-step method is presented to make it simple.

State the Null and Research Hypotheses Setting of hypotheses is the key element in any statistical test, parametric or non-parametric. It evolves from the research study design. The experimenter or researcher may have research questions to be answered by conducting a survey or experiment. The two types of hypotheses are null and alternate. The null hypothesis (usually denoted as H_0) is a statement that indicates no difference exists between conditions, groups or variables. The alternate hypothesis (H_1 or H_A), sometimes also called research hypothesis, is a statement differing from null hypothesis.

The alternative hypothesis may be directional, depending on the context of the research. Here we state some examples of null and its alternative hypotheses.

Example 6.3

H_0: Average fish yield in wetlands of two districts is the same.

H_A: Average fish yield of wetlands is different between two districts. For one-sided alternative, it is written as.

H_A: Average fish yield of wetlands of the district 1 is more than that in the district 2.

Example 6.4

H_0: Species richness between two sample sites is equal.

$H_{A:}$ Species richness between sample site 1 and sample site 2 is different, **or** species richness of sample site 1 is less than that of sample at site 2.

Choosing of Appropriate Test Statistic Test statistic is usually chosen on the basis of characteristic of data. The number of samples or groups, for instance, should be considered. Some tests are appropriate for two samples; some are suitable for three or more samples. Table 6.4 is provided for contextual reference for selection of suitable test statistics, albeit the list is not exhaustive to solve all kind of problems.

Measurement scale also plays a vital role in choosing test statistics. For example, non-parametric test is best suitable for data recorded in Likert scales, which are very often gathered during survey.

Compute the Test Statistic The test statistic value is computed by using the formula derived for particular type of test. This will be described later on in this chapter.

Determine the Critical Value For small samples, critical values corresponding to each non-parametric test procedures are available in any standard books. Reliable online sources can also be used. Each table provides a critical value to which we compare the computed test statistic. Finding a critical value requires a desired level of risk, the level of significance. In addition, other information like number of observations, groups and degrees of freedom is also required, depending upon the non-parametric test.

Compare the Obtained Value to the Critical Value Comparing the obtained value to the critical value helps to identify a difference based on a prefixed level of significance. Once this is done, one can state whether to reject the null hypothesis or not. Sometimes it is little bit confusing and unusual. In research, however, statement of results in terms of null hypothesis is standard practice. In the case of large samples, most of test statistic follows approximately normal distribution. In that case z-score is used to obtain critical value. For example, if we use a two-tailed test with 5% level of significance, we do not reject null hypothesis if absolute value of z is greater than 1.96. When the test statistic is carried out in software, it normally provides the value of the test statistic and p-value. A rule of thumb is that reject null hypothesis if the computed p-value is less than the desired level of significance.

Interpretation of Results At this stage, we can now give the meaning to the number and values obtained from the analysis, depending upon the context. If sample differences are observed, one can comment on the strength of those differences. Observed and expected results can also be compared.

6.5 Some Frequently Used Non-parametric Test Methods

There are large number of non-parametric methods to solve many real-life problems. However, this chapter describes most popular methods to solve specific problems.

This section describes non-parametric methods to solve the following four problems:

- Single sample problem to test population median
- Matched pair single sample problems. Alternatively, comparison of two related sample
- Two sample problems. Alternatively, comparison of two unrelated samples
- Multiple sample problems. Alternatively, comparison of three or more samples

6.5.1 Single Sample Problem

Here the basic objective is to test the known median of a population. Imagine that a random sample is drawn from a population of known median, say m_0. One wishes to make some inference on the median of the population based on the sample. The possible research questions are as follows:

- Whether the median of the sample (m) is equal to the theoretical value (m_0)
- Whether the median of the sample (m) is greater than the theoretical value (m_0)
- Whether the median of the sample (m) is less than the theoretical value (m_0)

The above research question can be stated in terms of hypotheses. The null hypothesis (H_0) may be written as H_0: $m = m_0$. The corresponding alternative hypotheses are as follows:

$$H_A : m \neq m_0; \text{different} \tag{6.1}$$

$$H_A : m > m_0; \text{greater than} \tag{6.2}$$

$$H_A : m < m_0; \text{less than} \tag{6.3}$$

The one sample **Wilcoxon signed rank** test is used to test the above hypothesis. This is an alternative to one-sample t-test. The required conditions for performing the sign test for the population median are less stringent than those for the t-test for the population mean. The sign test requires only that the sample data have been randomly selected. It is not required that the population be normally distributed. It should be noted, however, that the sign test is a hypothesis test for the population median, not the population mean. The key concept for performing the sign test for the median is that the data values are converted to either a plus sign (+) or a minus sign (−) by subtracting the median from the observed data values. If there is a preponderance of plus signs to minus signs, or vice versa (depending on the form of the hypothesis test), then this is evidence against the null hypothesis. Though the test can be performed on the basis of number of positive and negative sign (sign test or Wilcoxon sign test), it is dependent only on the sign. The value of the rank is ignored. However, we describe a better test, the Wilcoxon signed rank test, or Wilcoxon signed rank sum test, which includes both the sign and rank of the sample values.

Wilcoxon Signed Rank Test

In the Wilcoxon signed rank test, the samples are ordered and assigned rank based on the absolute values of the difference between the data value and the median (m_0). Then the Wilcoxon **T**, the test statistic for small samples, is computed by using the aforementioned positive and negative values as in the formula

$$T = \text{smaller of } \Sigma\text{R}_+ \text{ and } \Sigma\text{R}_-; \tag{6.4}$$

where PR_+ is the sum of all positive ranks and PR_- is the sum of all negative ranks. After the **Wilcoxon T** statistic is computed; it must be examined for significance. A table of critical values is available in any standard books (see Gibbons and Chakraborti 2020); it may also be available from reliable Internet sources. While consulting the table, choose the critical value depending on the alternative (one-tailed, two-tailed).

Decision Rule

Let T_{crit} be the critical value at predetermined level of significance and alternative hypothesis. Then reject the null hypothesis if $T < T_{\text{crit}}$. However, if the number of data values, *n*, exceeds those available from the table (normally $n > 25$), the large sample approximation can be performed. Then the test is performed by typical *z-test by* using the statistic

$$Z = \frac{T - \overline{X}_T}{S_T}; \text{where } \overline{X}_T = \frac{n(n+1)}{4} \text{ and } S_T = \sqrt{\frac{n(n+1)(2n+1)}{24}} \tag{6.5}$$

Example 6.5

In inland fisheries, this situation may arise in capture fisheries with selective gear. In gill net fisheries, for example, only target size based on the designed mesh size is captured. From the past experience, the median size of captured fish is a priori known. If mesh size is rightly working, one may expect that size of captured fish would be around that known median value. Suppose 12 fishes have been captured in gill net fishing, where targeted median length value was supposed to be 30 cm. The

Table 6.5 Computation sheet for Wilcoxon T

Length	$D_i = X_i - 30$	$\lvert d_i \rvert$	Rank	Signed rank
32	2	2	2	2
14	−16	16	10	−10
53	23	23	12	12
18	−12	12	7	−7
21	−9	9	5	−5
44	14	14	9	9
12	−18	18	11	−11
36	6	6	3	3
22	−8	8	4	−4
29	−1	1	1	−1
17	−13	13	8	−8
40	10	10	6	6

length (cm) data for 12 fishes were 31, 14, 53, 18, 21, 44, 12, 36, 22, 29, 17 and 40. The researcher was interested to examine whether the median capture size differs from 30 cm or not.

Solution
Step 1: Setting Null Hypothesis

H_0: *Whether the median length of the fish is 30 cm.*
H_1: *Median length of fish differs from 30 cm.*

Step 2: Computation of Test Statistic *The sum of positive ranks* $PR_+ = 32$ *and the sum of the negative ranks* $PR_- = 46$. *Therefore, the Wilcoxon test statistic* $T =$ *smaller of* PR_+ *and* $PR_- = 32$ (see Table 6.5).

Step 3: Obtaining Critical Value *To obtain critical value, it is very important to fix the level of significance. Here in this example, we set the level of significance as 5%. The p-value to be considered in the statistical tables is 0.05. For two- sided test, it is 0.025. Thus the critical values obtained here are as follows:*

- 14 for two-tailed (median is not equal to 30 cm)
- 17 for left- tailed (median is less than 30 cm)
- 64 for right- tailed (median is greater than 30 cm).

Step 4: Interpretation *Since $T = 32$ is greater than 14, we cannot reject the null hypothesis at 5% level of significance. Alternatively, we conclude that, at 5% level of significance, there is no evidence that median size differs from 30 cm.*

6.5.2 Comparing Two Dependent Samples

This problem arises when data is recorded for two observed variables on the same subject. We call this as matched pair samples. For example, a researcher is interested on the certain parasite infection on fish. The researcher suspected that the infection of parasite might differ from body to the head. The parasite infections at the head and body are two values recorded on the same fish. This is a case of matched pair samples.

In parametric case, paired *t*-test is usually used for paired sample problem. The non- parametric test used to solve the problem is again the 'Wilcoxon signed rank test', which has been described above. Here the hypotheses are stated as

H_0: Medians of the two populations are equal.
H_1: Medians of the two populations are not equal.
H_1: Median of population 1 is greater than that of population 2.
H_1: Median of population 1 is less than that of population 2.

The above hypotheses can also be written in mathematical notation. Let m_1 and m_2 be the two medians of population 1 and population 2, respectively. The above mentioned hypotheses are then can be written as

$$H_0 : m_1 = m_2$$

$$H_1 : m_1 \neq m$$

$$H_1 : m_1 > m_2$$

$$H_1 : m_1 < m_2$$

The computation of **WilcoxonT** statistic is similar to that of single sample problem of testing known median value. The only difference is that instead of using values of difference about median, pairwise difference is used to compute the **Wilcoxon T** statistic. Thus the expected median of the derived pairwise difference is zero. When pairwise difference is obtained, the problem resembles to the one sample problem with known median of zero value. We demonstrate this case in the following example:

Example 6.6

Table 6.6 shows the parasite infection to a fish species. Thirteen fishes were examined for the number of parasites present in two body sectors. Sector-wise parasite infection is recorded in Table 6.6. It is hypothesized that parasite infection on two sectors is different. This is a two dependent sample problem, as parasite infection in sector 1 and sector 2 is recorded on the same fish.

Table 6.6 Parasite infection in two body sectors of fish

Fish ID	Sector 1	Sector 2
1	1	1
2	0	2
3	0	2
4	0	0
5	2	3
6	0	4
7	1	0
8	1	2
9	1	3
10	2	0
11	0	4
12	0	2
13	0	2

Solution

Step 1: Stating of Hypothesis *The null hypothesis here is that the median parasite infection of sector 1 is equal to that of sector 2.*

The alternative hypothesis is that there is difference of median parasite infection between sector 1 and sector 2. It is a two-sided alternative hypothesis.

Let us denote that m_1 *is the median parasite infection of sector 1 and* m_2 *is the parasite infection of sector 2. Then the mathematical representations of the null and alternative hypotheses are as*

$$H_0 : m_1 = m_2$$

$$H_1 : m_1 \neq m_2$$

Step 2: Computation Test Statistic *From* Table 6.7, *it is found that* $T_+ =$ *the sum of positive ranks* $= 8.5$; $T_- =$ *the sum of negative ranks* $= 57.5$; $T =$ *smaller of* T_+*; and* $T_- = 8.5$.

Step 3: Obtaining Critical Value *The table for critical values of Wilcoxon's signed rank test has been consulted. Here the alternative hypothesis is two sided. So we have chosen the critical value for sample size 13 and for two-tailed test. The critical values for 5% and 1% level of significance are 17 and 9, respectively.*

Step 4: Interpretation *The computed T value is 8.5. Since* $T = 8.5$ *is less than 9, the null hypothesis is rejected at 1% level of significance. Obviously, the null hypothesis is rejected at 5% level of significance. Hence we may conclude that there is evidence of significant difference (1% level of significance) of parasite infection between sector 1 and sector 2 of the body of the fish.*

Table 6.7 Computation of test statistics for paired sample test. S_1: sector 1; S_2: sector 2

Fish ID	*S*1	*S*2	*S*1 – *S*2	|*S*1 – *S*2|	Rank	Signed rank
1	1	1	0	0	Excluded	
2	0	2	−2	2	6.5	−6.5
3	0	2	−2	2	6.5	−6.5
4	0	0	0	0	Excluded	
5	2	3	−1	1	2	−2
6	0	4	−4	4	10.5	−10.5
7	1	0	1	1	2	2
8	1	2	−1	1	2	−2
9	1	3	−2	2	6.5	−6.5
10	2	0	2	2	6.5	6.5
11	0	4	−4	4	10.5	−10.5
12	0	2	−2	2	6.5	−6.5
13	0	2	−2	2	6.5	−6.5

6.5.3 Comparing Two Independent Samples: The Mann-Whitney *U*-Test

In parametric testing of hypothesis, two independent samples are compared on the basis of two means of the populations. In most of the cases, the variance of the two populations is usually assumed to be the same. The popular *t-test* is used to perform the test of equality of two means. The non-parametric alternative, however, does not assume any equality of variances of the two populations. Alternatively, the equality of two medians of the populations is tested in the non-parametric procedure.

The **Mann-Whitney *U*-test** is a non-parametric procedure to address the above problem. Specifically, it is used to compare two independent or not related samples. The Wilcoxon rank sum test is a similar non-parametric test to the **Mann-Whitney *U*-test**. We will describe how to perform and interpret the **Mann-Whitney *U*-test**.

For computing the **Mann-Whitney *U*-test** statistic, data values recorded for two samples are combined. The combined values are then ordered and assigned rank. The logic behind the derivation of U-statistic is that if the values from the two samples are randomly mixed in the rank ordering or if they are clustered at the opposite end when combined. A random rank order would indicate that the two samples are not different, while a cluster of one sample's values would indicate a difference between them. Suppose there are n_1 and n_2 observations for sample 1 and sample 2, respectively. For each of the samples, we compute ***U*-test** statistic using the following formula:

$$U_i = n_1 n_2 + \frac{n_i(n_i + 1)}{2} - \sum_{i=1}^{2} R_i \tag{6.6}$$

where U_i is the test statistic for the *i*th sample ($i = 1, 2$) and $\sum_{i=1}^{2} R_i$ is the sum of ranks from the *i*th sample. The final value of **Mann-Whitney *U*-test** statistic is the

Table 6.8 Fish yield of 23 beels in two districts

Beel	District	Yield (kg/ha)	Rank
1	Morigaon	46	1
2	Morigaon	62.5	2
3	Morigaon	85	3
4	Morigaon	142	4
5	Morigaon	170	5
6	Morigaon	196	6
7	Morigaon	220	7
8	Morigaon	250	8
9	Morigaon	256	10
10	Morigaon	340	13
11	Morigaon	363	14
12	Morigaon	409	15
13	Morigaon	416	16
14	Morigaon	432	19
15	Morigaon	488	22
1	Nagaon	250	9
2	Nagaon	285.7	11
3	Nagaon	290	12
4	Nagaon	421	17
5	Nagaon	429	18
6	Nagaon	445	20
7	Nagaon	446	21
8	Nagaon	493	23

smaller of the U_1 and U_2. After the U-test statistic is computed, this is examined for significance. We may find the critical value from the table of critical values for U-statistic, which is available in any standard books or in reliable Internet sources. However, if the number of observations exceeds the number available in the table, the large sample approximation may be used. The *z-test* is then used, where the z-score is computed by using the following formula:

$$Z = \frac{U - \bar{x}_U}{S_U}, \text{ where } \bar{x}_U = \frac{n_1 n_2}{2} \text{ and } S_U = \sqrt{\frac{n_1 n_2 (n_1 + n_2 + 1)}{12}} \tag{6.7}$$

Example 6.7

A small survey was conducted to compare the fish yield between two districts. Fifteen wetlands from Morigaon district and eight wetlands from Nagaon district were selected randomly, and corresponding yields were recorded as shown in Table 6.8. The objective was to compare the yields between Morigaon and Nagaon districts.

Solution
Step 1: Stating of hypothesis

H_0: *Median fish yield of floodplain wetlands in Morigaon and Nagaon districts is equal.*
H_1:
Median fish yield of floodplain wetlands in Morigaon and Nagaon district is different.

Step 2: Computation of Mann-Whitney U-Statistic *After combining the values of fish yield, all the values were arranged, and ranks are assigned in ascending order. The rank received by each beel is shown in* Table 6.8. *The sum of ranks of Morigaon district* = $R_M = 145$, *and the sum of ranks of Nagaon district* $R_N = 131$. *The U-statistic for Morigaon district* = $U_M = 15 \times 8 + (15 \times (15 + 1))/2–145 = 95$. *Similarly, the U- statistic for Nagaon district* = $U_N =$

$$15 \times 8 + 8 \times (8 + 1)/2 - 131 = 25. U = \text{smaller of } U_M \text{ and } U_N = 25.$$

Step 3: Determining the Critical Value *The critical value at 5% level of significance for two-tailed test corresponding to sample sizes 8 and 15 is obtained from the table as 91.*

Step 4: Interpret Results *Since* $U = 25$ *is less than 91, the critical value, hence we reject the null hypothesis. Alternatively, we can conclude there is a significant difference between yields of Morigaon and Nagaon district.*

6.5.4 Comparing Three or More Independent Samples: The Kruskal- Wallis Test

The **Kruskal–Wallis** test is a non-parametric procedure for testing whether the samples originate from the same distribution. More specifically, it is used to compare two or more independent samples of equal or different sample sizes. It extends the Mann–Whitney *U*-test. The parametric alternative is **one-way ANOVA**. The **Kruskal–Wallis *H*-test** statistic is used to compare more than two independent samples. The null hypothesis is stated in terms of the population. Since it is a non-parametric method, the Kruskal–Wallis test does not assume a normal distribution of the residuals, unlike the analogous one-way analysis of variance. If the researcher can make the less stringent assumptions of an identically shaped and scaled distribution for all groups, except for any difference in medians, then the null hypothesis is that the medians of all groups are equal, and the alternative hypothesis is that at least one population median of one group is different from the population median of at least one other group.

When Kruskal–Wallis test leads to significant results, then at least one of the samples is different from the other samples. However, the test does not identify where the difference occurs. Moreover, it does not identify how many differences occur. To identify the specific difference occurring between two pairs, one may analyse specific sample pairs for significant differences. The Mann–Whitney *U*-test is a useful method for performing sample contrasts between individual samples test.

Method of Computation of Kruskal–Wallis *H*-Test Statistic
Suppose there are *six* groups of independent samples each of size $n_i, = 1, 2, \ldots, g$ with total number of observations *N*. Rank all data values from all groups together; i.e. rank the data from 1 to *N* ignoring the group membership. Assign any tied values the average of the ranks they would have received had they not been tied. The test statistic is given by

$$H = \frac{12}{N(N+1)} \sum_{i=1}^{g} n_i \bar{r}_i^2 - 3(N+1) \tag{6.8}$$

where $\bar{r}_i =$ is the average rank of all observations in group *i*. In case of ties, *H* is divided by a correction factor

$$1 - \frac{\sum_{i=1}^{G} (t^3 - t_i)}{N^3 - N}$$

where *G* is the number of groupings of different tied ranks and t_i is the number of tied values within group *i* that are tied at a particular value. This correction usually makes little difference in the value of *H* unless there are a large number of ties. Finally, the computed *H* is compared with the critical value to reject or accept the null hypothesis. For small sample, the critical value is obtained from table of critical values for *H*-test. The table is mostly available for three groups, and the number of observations in each group is less than or equal to 5. Alternatively, chi-square test with $g - 1$ degree of freedom can be used. Suppose the critical value (from Table *H-test* or from table of chi-square test) for certain level of significance (say 5%) obtained is H_{crit}. When the null hypothesis of equality of population distributions is false, then the value of *H* is expected to be larger. Therefore, reject the null hypothesis if $H > H_{\text{crit}}$.

Example 6.8
In a wetland survey, there were three different types of management, viz. unmanaged, well managed and moderately managed. A total of 60 wetlands were surveyed and yield of each wetland was recorded. Out of 60 wetlands 14, 19 and 27 wetlands were in the category of 'unmanaged', 'well managed' and 'moderately managed', respectively. The research question was whether the fish yields of wetlands under different management categories are different or not.

Solution
Step 1: Stating Hypothesis *Research question: whether the fish yields of wetlands under different management categories are different or not?*

Table 6.9 Ranks of yield of surveyed wetland for different management categories

Sl no.	Unmanaged	Well managed	Moderately managed
1	1	13	4
2	2	22	5
3	3	27	8
4	6	29	9
5	7	34	10
6	11	39	12
7	14	42	18
8	15	44	19
9	16	45	20
10	17	46	21
11	24	47	23
12	40	49	25
13	41	51	26
14	54	53	28
15	–	55	30
16	–	56	31
17	–	57	32
18	–	59	33
19	–	60	35
20	–	–	36
21	–	–	37
22	–	–	38
23	–	–	43
24	–	–	48
25	–	–	50
26	–	–	52
27	–	–	58
Average	17.93	43.58	27.81

H_0: *The fish yield distributions of three management categories are equal.*
H_1: The *yield distribution of at least one management category differs.*

Step 2: Computation of Kruskal–Wallis *H* Statistic

1. Ranking of yield of combined samples has been done. The ranks of each of the management category are recorded in Table 6.9. Note that average rank of three categories is shown in Table 6.9.
2. Thus, *H* is computed as follows:

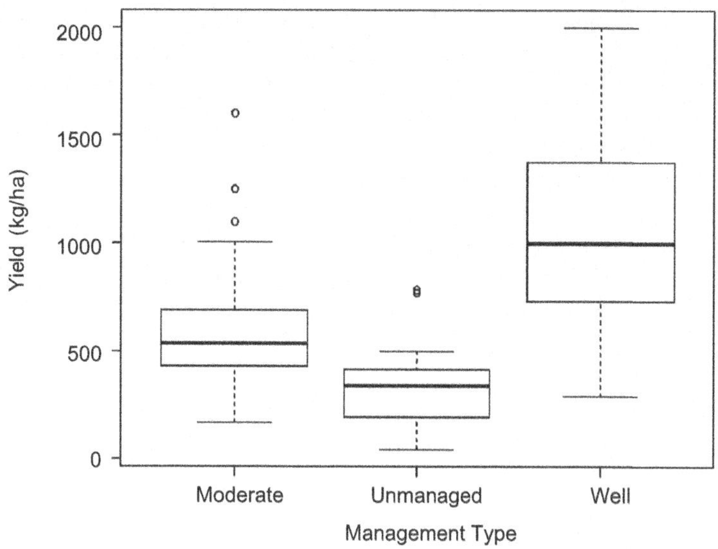

Fig. 6.2 Yield distribution in different management categories

$$H = \frac{12}{60(60+1)}\left(14 \times 17.928^2 + 19 \times 43.58^2 + 27 \times 27.814^2\right) - 3 \times (60+1)$$
$$= 18.21$$

Step 3: Comparing Critical Value *Here the degrees of freedom* $= 3 - 1 = 2$. *After consulting chi-square table, we obtain the critical value, for* $df = 2$ *and level of significance* $= 0.05$, *as* 5.99. *Since* $H = 18 > 5.99$, *we reject the null hypothesis.*

Step 4: Interpretation *In the above step, the null hypothesis has been rejected. So we may conclude that there are sufficient evidence that fish yield of at least one management category is significantly different. It is further required to identify which of the management categories are different, as far as fish yield is concerned. One objective way is to perform Mann–Whitney U-test for pairwise comparison. Though subjective, the same can be inspected visually by boxplot as shown in* Fig. 6.2. *From* Fig. 6.2, *it is evident that fish yield of all the three management categories is different.*

6.6 Conclusion

This chapter describes the concept of non-parametric test for solving some statistical problems. However, the test procedures are not exhaustive. Only the basic non-parametric test procedures, which are very often required for data analysis, have been discussed. The emphasis has been given on the systematic execution of the test procedures. Derivation of the formula has been avoided. The most important to implement these procedures is the identification of statistical problem and formulation of research question. Subsequently, it requires the formulation and stating of prudent hypotheses to find out the answer to the research question. In most of the cases, computation can be done manually, which is the biggest strength of the methods described in this chapter, albeit software is available to carry out those test procedure. The advantage of using software is that it provides p- value. So, it becomes very easy to reject or accept the null hypothesis with little knowledge of statistics. There are commercial (SPSS, SAS, MATLAB, STATA, etc.) and free R softwares available to implement those procedures. However, those simple computations can easily be done in Excel.

Exercise

1. Under what circumstances would you use a non-parametric test?
   (a) When your data does not meet the assumptions for a parametric test
   (b) In a pilot study
   (c) When you think your sample size is too big
   (d) When you do not really understand a parametric test
2. If you achieved a *p*-value of 0.04 on a two-tailed test, what would the equivalent one- tailed *p*-value be?
   (a) 0.02
   (b) 0.002
   (c) 0.2
   (d) 0.06.
3. You have carried out a Kruskal–Wallis test. There are significant differences between the three groups you are testing. How might you conduct your pairwise comparisons?
   (a) Use a *t*-test
   (b) Use the Mann–Whitney test
   (c) Use the Wilcoxon test
   (d) None of the above. Post hoc analyses cannot be carried out with non-parametric data
4. What is the non-parametric alternative to *z*-test or *t*-test for single sample?
   (a) Kruskal–Wallis test
   (b) Mann–Whitney *U*-test
   (c) Wilcoxon signed rank test
   (d) Run test
5. What is the non-parametric alternative to paired *t*-test?
   (a) Wilcoxon signed rank test
   (b) Mann–Whitney *U*-test
   (c) Run test
   (d) Kruskal–Wallis test
6. What is the non-parametric alternative to one-way ANOVA?

(a) Wilcoxon signed rank test
(b) Kruskal–Wallis test
(c) Run test
(d) Mann–Whitney U-test

7. Non-parametric tests make certain assumptions about
   (a) Population size
   (b) The underlying distribution
   (c) Sample size
   (d) None of the above
8. The Wilcoxon signed rank test tends to z-test when sample size is
   (a) More than 25.
   (b) Less than 10.
   (c) Equal to 15.
   (d) None of the above
9. Large sample approximation of the Mann–Whitney U-test is the
   (a) z-test
   (b) t-test
   (c) Kruskal–Wallis test
   (d) F-test
10. To apply the Mann–Whitney U-test to determine whether the location of population 1 is different from the location of population 2, the samples must be:
    (a) Drawn from normal populations
    (b) Drawn from matched pairs experiment
    (c) Independent
    (d) Larger than 30
11. To apply Mann–Whitney U-test to compare two independent samples, which of the following statements is true?
    (a) Variance of the two population is the same
    (b) Variance of the two population is known
    (c) Variance of the two population is not equal
    (d) Information on the variance is not required
12. In a single sample Wilcoxon signed rank test, what population parameter is tested?
    (a) Mean
    (b) Median
    (c) Variance
    (d) Mode
13. The non-parametric hypothesis for comparing two samples is
    (a) Medians of the two population are equal
    (b) Means of the two population are equal
    (c) Variances of the two population are equal
    (d) Mode of the two population is equal
14. The number of species counts at 10 sites in a river stretch are 44, 27, 11, 8, 15, 6, 22, 10, 12 and 69. What is the rank of 10 to perform a non-parametric test?

(a) 1
(b) 10
(c) 2
(d) 3

15. The number of species counts in 10 sites in a river stretch are 44, 27, 11, 8, 15, 6, 22, 10, 11 and 69. What is the rank of 11 to perform a non-parametric test?
(a) 3
(b) 4.5
(c) 6
(d) 4
16. The annual fish yield (kg/ha) of five wetlands are 46, 170, 250, 429 and 300. What are the mean and variance of rank of above values?
(a) Mean = 2; variance = 1
(b) Mean = 1; variance = 1
(c) Mean = 3; variance = 2.5
(d) None of the above
17. The species richness at 10 sites is 45, 28, 12, 9, 16, 7, 23, 11, 12 and 70 in a river stretch. A researcher wishes to test whether species richness in the stretch is more than 15 or not. What non-parametric test is suitable here?
(a) Run test
(b) Wilcoxon signed rank test
(c) Mann–Whitney U-test
(d) Kruskal–Wallis test
18. The species richness at 10 sites is 45, 28, 12, 9, 16, 7, 23, 11, 12 and 70 in a river stretch. A researcher wishes to test whether species richness in the stretch is more than 19 or not. Which of the following types of test is suitable?
(a) One-tailed test
(b) Two- tailed test
19. A fisherman uses two types of gillnet of two different sizes to capture fish in a fishing zone. A researcher randomly selects 15 and 12 fishes captured by gillnet 1 and 2, respectively. Which test is suitable to test that size at capture by two gillnets is equal?
(a) Mann–Whitney U-test
(b) Kruskal–Wallis test
(c) Wilcoxon signed rank test
(d) None of the above
20. In a river, the species richness is as follows: 20, 10 and 15 at 3 sites within the stretch 1; 30, 20, 25, 22 and 10 at 5 sites within the stretch 2; and 22, 12, 17 and 13 at 4 sites within the stretch 3. Which non-parametric test is suitable to test equality of species richness in the three stretches of the river?
(a) Run test
(b) Mann–Whitney U-test
(c) Kruskal–Wallis
(d) Wilcoxon signed rank test

Further Reading

Biradar RS (2002) Fisheries statistics, 2nd edn. Central Institute of Fisheries Education, Mumbai

Sheskin DJ (2000) Handbook of parametric and nonparametric statistical procedures, 2nd edn. Chapman & Hall/CRC

Gibbons JD, Chakraborti S (2020) Nonparametric statistical inference, 4th edn. Marcel Dekker, Inc, New York

Hesse CA, Ofosu JB, Nortey EN (2017) Introduction to nonparametric statistical methods. Akrong Publications Ltd., Accra

Analysis of Variance (ANOVA) and Design of Experiments

7

Key Topics

- Concept of ANOVA
- One-way ANOVA
- Basic concept of design of experiment
- Two-way ANOVA

Many experiments in fisheries science involve more than two treatment groups. We must not use *t*-test to compare more than two groups because multiple *t*-tests performed on the same data set increase likelihood of falsely reporting a significant difference. For testing the significance of means of several independent groups of observations, a powerful parametric statistical approach is analysis of variance (ANOVA). It was developed by one of the founders of modern statistical theory, Sir Ronald Fisher, in the 1920s. It assesses potential differences in a continuous-level (interval/ratio) dependent variable by a categorical-level (nominal/ordinal) independent variable having two or more categories. For example, an ANOVA can examine potential differences in fish yield of beels by stocking density or macrophyte infestation (low vs. moderate vs. high). The two main aims of classical ANOVA are (1) to examine the relative contribution of different sources of variation (factors or combination of factors, i.e. predictor variables) to the total amount of variability in the response variable and (2) to test the null hypothesis that population groups or treatment means are equal. In this chapter, assumptions of ANOVA, one-way ANOVA, two-way ANOVA, post hoc tests, repeated measures ANOVA and design of experiments with examples on inland fisheries will be discussed. The computations are involved in partitioning of total amount of variability in response variable into explained (treatments or factors or combination of factors) and unexplained variability (residual). Excel/SPSS has been demonstrated to carrying out ANOVA.

B. K. Das et al., *Concept Building in Fisheries Data Analysis*,
https://doi.org/10.1007/978-981-19-4411-6_7

7.1 Assumptions of ANOVA

1. The population from which samples are drawn is normally distributed.
2. Independence of cases: the samples are independent of each other.
3. Homogeneity of variances: the variances among the groups are approximately equal.

These assumptions can be tested using any statistical software (like SPSS/SAS/R). Normality of the distribution of the data under study can be tested using histograms, the values of skewness and kurtosis or tests such as Shapiro–Wilk or Kolmogorov–Smirnov. The assumption of independence can be determined from the design of the study. It is important to note that ANOVA is not robust to violations to the assumption of independence. This is to say that even if you violate the assumptions of homogeneity or normality, you can conduct the test and basically trust the findings. However, the results of the ANOVA are invalid if the independence assumption is violated. The assumption of homogeneity of variance can be tested using Levene's test. In general, with violations of homogeneity, the analysis is considered robust if you have equally sized groups. With violations of normality, continuing with the ANOVA is generally performed, if you have a large sample size.

7.2 Hypotheses and Procedure of ANOVA

H_0: There is no significant difference among the groups, that is,

$$\mu_1 = \mu_2 = \ldots = \mu_n$$

H_1: There is at least one significant difference among the groups, that is,

$$\mu_1 \neq \mu_2 \neq \ldots \neq \mu_n$$

After data cleaning, the investigator must test the assumptions of ANOVA. They must then calculate the F-ratio and the associated probability value (p-value). In general, if the p-value associated with the F is smaller than 0.05, then the null hypothesis is rejected and the alternative hypothesis is supported. The rejection of null hypothesis concludes that the means of all the groups are not equal. At this point you could run post hoc tests which are t-tests examining mean differences between the groups. There are several multiple comparison tests that can be conducted that will control for Type I error rate, including the Duncan, Bonferroni, Scheffe, Dunnett and Tukey tests. Such tests are never to be used or interpreted when the F-test does not show the existence of significant differences. Post hoc tests tell the researcher which group means are different from each other.

7.3 One-Way ANOVA

One-way ANOVA has just one independent variable, i.e. only one factor is involved. For example, difference in fish production of beels can be assessed by different stocking densities, e.g. low, medium and high categories to compare. It examines question like 'Are there differences in fish production of beels by stocking density (no vs. moderate vs. high)'. A typical one-way ANOVA table for a factor at v levels each replicated n_i times such that $\sum n_i = n$ is as follows (Table 7.1):

Example 7.1: Illustration

We wish to know if there is a difference in fish yield of three beels practising supplementary stocking (no/ moderate/ high) over a period of 5 years. Here, we have one fixed factor at three levels (1 = control, 2 = moderate and 3 = high stocking rate) and five replicates for each level of factor variable. The data is presented in Table 7.2.

Step 1: State the Null Hypotheses

H_0: No effect of stocking rate on yield of beels, i.e. $\mu_1 = \mu_2 = \mu_3$.

H_1: At least one of the stocking rates has significant effect on yield of beels, i.e., $\mu_1 \neq \mu_2 \neq \mu_3$.

Step 2: Compute the Test Statistic

Step-by-step procedure for solving this data using SPSS is as follows:

Feed the data into SPSS. Label the factor variable as Stocking, and code the levels of factor variable as 1 for 'no stocking', 2 for 'moderate stocking' and 3 for 'high stocking'. The response variable, i.e. fish yield, is labelled as variable.

Yield. Go to Click Analyze, Compare Means and **One-Way ANOVA...**

Table 7.1 A typical format of one-way ANOVA

Source of variation	DF[a]	SS[b]	MSS[c]	^{F}ratio
Between treatments	v-1	SST	MST = SST/v-1	$\frac{MST}{MSE} \sim F_{(v-1),(n-v)}$
Within treatments (error)	n-v	SSE	MSE = SSE/n-v	
Total	n-1	TSS		

[a]Degrees of freedom
[b]Sum of square
[c]Mean sum of square

Table 7.2 Fish yield for different levels of stocking

Stocking rate	Rep. 1	Rep. 2	Rep. 3	Rep. 4	Rep. 5
No stocking	717	405	398	367	300
Moderate	719	800	1225	720	773
High	1329	1501	1395	1300	1465

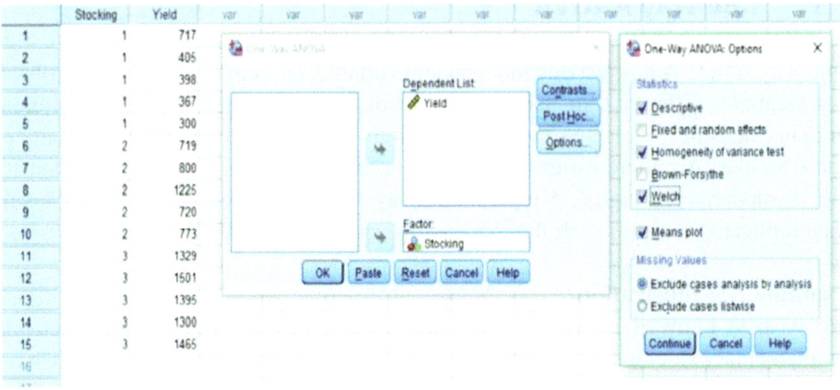

In **One-Way ANOVA** dialogue box, send the dependent variable, Yield, into the **Dependent List** box, and transfer the factor variable, Stocking, into the Factor box by using the arrow button or by dragging and dropping the variables into the respective boxes.

Click on the **Options**.. . button. **One-Way ANOVA: Options** dialogue box opens. Check the boxes for Descriptive and Homogeneity of variance test. Click the **Continue** button to return to the **One-Way ANOVA** dialogue box. Click the **Post Hoc** button. **One- Way ANOVA: Post Hoc Multiple Comparisons** dialogue box opens up; select Duncan under Equal Variances Assumed. Click the **Continue** button to return to the One-Way ANOVA dialogue box.

Click the **OK** button to generate the output.

Step 3: Statistical Decision

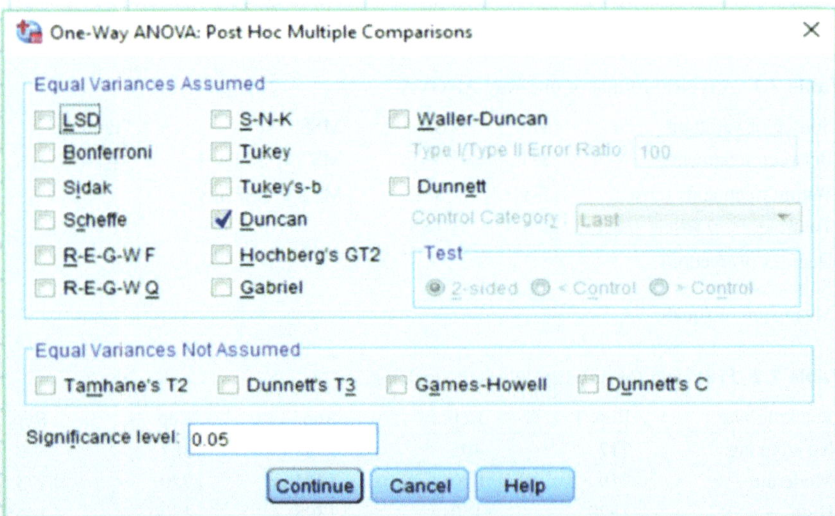

Table 7.3 ANOVA table for effect of factor stocking on yield of beels

Source	Sum of squares	df	Mean square	*F*	Sig.
Between groups	2323354.533	2	1161677.267	43.952	0.000
Within groups	317166.400	12	26430.533		
Total	2640520.933	14			

Here, the calculated *p-value* of the *F*-statistic is very less shown as 0.000 under sig. Column of Table 7.3. Hence, it is highly significant, and we reject the null hypothesis and conclude that there is significance of stocking rates on the yield of beels, and beels with high stocking rates have more yield.

Step 4: Interpret Results

ANOVA indicated statistically significant ($p < 0.05$) *effect of stocking on the fish yield rate that was the lowest in unstocked beels with an average of* 437 ± 73.32 kg ha^{-1}. *Among the stocked beels, average fish yield rates were higher in beel that was stocked at higher density* (1398 ± 39 kg ha^{-1}) *than that in the beel, which was stocked at lower density (average* 847 ± 95 kg ha^{-1}*).*

7.4 Two-Way ANOVA

It refers to an ANOVA using two independent variables, i.e. two factors are involved (Table 7.4).

For example, a two-way ANOVA can examine differences in fish production of beels (the dependent variable) by macrophyte infestation (low vs moderate vs high) and stocking status (stocked vs unstocked). It can be used to examine the interaction between the two categorical variables. Interactions indicate that differences are not uniform across all categories of the categorical variables. For example, stocked beels may have high production overall compared to unstocked beels, but this difference could be greater (or less) in low macrophyte infestation compared to high macrophyte infestation. It is also called as factorial ANOVA. A typical two-way ANOVA table for one factor (A) at v levels and other factor (B) at b levels each replicated r times such that $r \times v \times b = n$ is as shown in Table 7.5.

Table 7.4 Duncan post hoc analysis. Subset for alpha =0.05

Stocking	1	2	3
No stocking	437.40		
Moderate stocking		847.40	
High stocking			1398.00
Sig.	1.000	1.000	1.000

Table 7.5 A typical format of one-way ANOVA

Source of variation	DF[a]	*SS*[b]	*MSS*[c]	^{F}ratio
Factor A	*v*-1	SSA	MSA	$\frac{\text{MST}}{\text{MSE}} \sim F_{(v-1),vb(r-1)}$
Factor B	*b*-1	SSB	MSB	$\frac{\text{MSAB}}{\text{MSE}} \sim F_{(v-1)(b-1),vb(r-1)}$
Interaction (A × B)	(*v*-1) (*b*-1)	SS(A × B)	MSAB	$\frac{\text{MSAB}}{\text{MSE}} \sim F_{(v-1)(b-1),vb(r-1)}$
Error total	*vb*(*r*-1) *rvb*-1	SSE TSS	MSE	

[a]Degrees of freedom
[b]Sum of square
[c]Mean sum of square

Table 7.6 Fish yield for different stocking status and macrophyte infestation

		Macrophyte infestation	
Stocking status	Low	Moderate	High
Stocked	1450	657.9	729
	1333	428	432
	1250	708	500
	1121	488	214
Unstocked	147	507	41
	363	782	85
	256	409	62
	416	500	46

Example 7.2: Illustration

We wish to know if there is a difference in fish yield of stocked and unstocked beels with macrophyte infestation. The data is presented in Table 7.6. The two factors are stocking status and macrophyte infestation. The factor macrophyte infestation has three levels (low/ moderate/ high), whereas stocking status has two levels (stocked/ unstocked).

Step 1: State the Null Hypotheses

H_0: No effect of stocking status on yield of beels.

H_0: No effect of macrophyte infestation on yield of beels.

H_0: No interaction between macrophyte infestation and stocking status on the yield of beels.

Step 2: Statistical Test

Since two factors are involved, we need to use a two-way ANOVA, with possibly a multiple comparison post hoc test, if necessary.

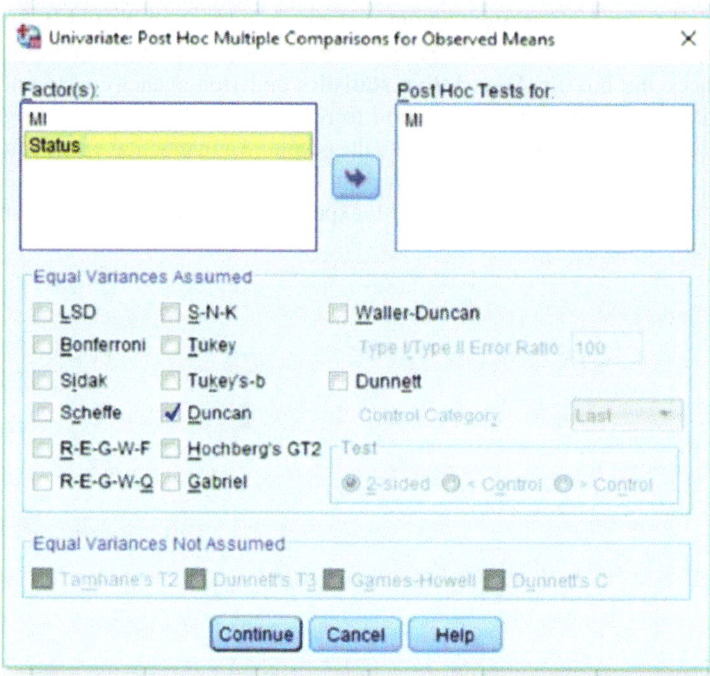

Step 3: Compute the Test Statistic

Step-by-step procedure for solving this data using SPSS is as follows:

Feed the data into SPSS. Label one factor variable as Macrophyte infestation, and code the levels of variable as 1 for 'low', 2 for 'moderate' and 3 for 'high'. Label the other factor variable as Stocking status, and code the levels of factor variable as 1 for 'stocked' and 2 for 'unstocked'. The response variable, fish yield, is labelled as the variable Yield. Go to Click **Analyze** ↑ **General Linear Model** ↑ **Univariate...** In **Univariate** dialogue box, send the dependent variable, Yield, into the **Dependent Variable:** box, and transfer the factor variable, Stocking, into the **Fixed Factor(s):** box by using the arrow button or by dragging and dropping the variables into the respective boxes. Click on the **Plots...** button. **Univariate: Profile Plots** dialogue box opens. Send the independent variable, Macrophyte Infestation, from the **Factors:** box into the **Horizontal Axis:** box, and transfer the other independent variable, Stocking status, into the **Separate Lines:** box. Click the Add button. **MI*status** has been added to the Plots: box. Click the Continue button to return to the Univariate dialogue box. Click the **Post Hoc...** button. **Univariate:**

Post Hoc Multiple Comparisons for Observed Means dialogue box opens up. Send MI (Macrophyte Infestation) from the Factor(s): box to the Post Hoc Tests for: box. Then check the box for Duncan under Equal Variances Assumed—area. Click the Continue button to return to the **Univariate** dialogue box. Click the Options... button. **Univariate:** Options dialogue box opens up. Send MI, Status and MI*status from the Factor(s) and Factor Interactions: box to the Display Means for: box. Check the box for **Descriptive statistics** and Homogeneity tests under Display—area. Click the **Continue** button to return to the **Univariate** dialogue box. Finally, click the **OK** button to generate the output. An interaction effect can usually be seen as a set of non-parallel lines. We can see from this graph that the lines do not appear to be parallel, and we might expect the interaction to be statistically significant.

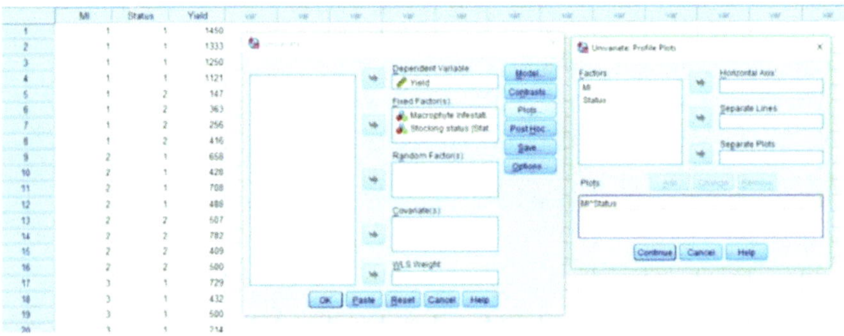

Step 4: Statistical Decision

You can see from the '**Sig.**' column (see Table 7.7) that we have a statistically significant interaction at the $p = 0.001$ level. We can see from the table above that there was statistically significant difference in mean yield between status ($p = 0.001$) and ($p = 0.001$). Hence, it is highly significant, and we reject the null hypothesis and conclude that the stocking status, macrophyte infestation and interaction have significant effect on the yield of beels.

Table 7.7 Results of two-way ANOVA

Source	Sum of squares	DF	MSE	*F*	Sig.
Corrected model	3431661.954	5	3431661.954	33.570	0.000
Intercept	6961835.884	1	6961835.884	340.519	0.000
MI	1121899.080	2	560949.540	27.437	0.000
Status	1352277.900	1	1352277.900	66.143	0.000
MI × status	957484.973	2	478742.487	23.416	0.000
Error	368006.413	18	20444.801		
Total	10761504.250	24			
Corrected total	3799668.366	23			

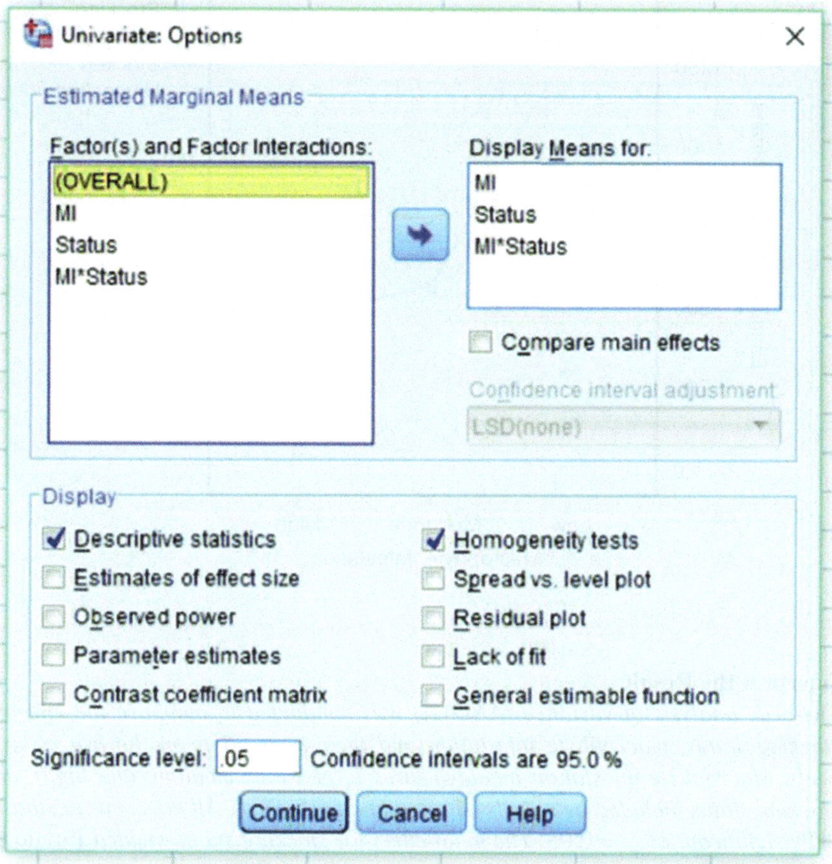
Univariate: Options
Estimated Marginal Means
Factor(s) and Factor Interactions:
(OVERALL)
MI
Status
MI*Status
Display Means for:
MI
Status
MI*Status
Compare main effects
Confidence interval adjustment
LSD(none)
Display
Descriptive statistics
Estimates of effect size
Observed power
Parameter estimates
Contrast coefficient matrix
Homogeneity tests
Spread vs. level plot
Residual plot
Lack of fit
General estimable function
Significance level: .05
Confidence intervals are 95.0 %
Continue
Cancel
Help

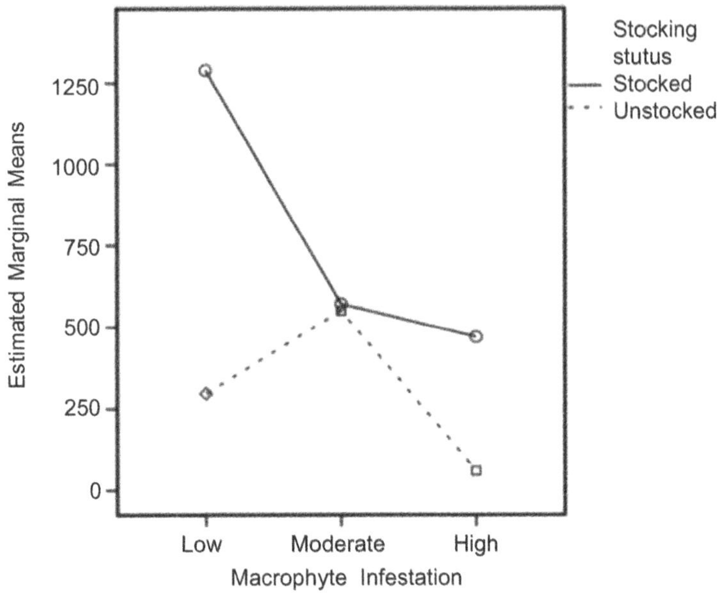

Interpret the Result

Two-way analysis of variance (ANOVA) was conducted to compare the effect of stocking status, macrophyte infestation and their interaction on the fish yield of beels. Macrophyte infestation included three levels (low, medium, and high), and stocking status included two levels (stocked and unstocked). All effects were statistically significant at ¯ = 0.05. *The main effect for stocking status yielded* F-ratio (1, 18) = 66.14, p = 0.001, *indicating a significant difference between stocked* (776 ± 118 kg/ha) *and unstocked beels* (301 ± 68 kg/ha). *The main effect for MI yielded an* F-ratio (2, 18) = 27.44, p = 0.001 *indicating a significant difference between low macrophyte-infested beels* (264 ± 92 kg/ha), *medium macrophyte-infested beels* (560 ± 49 kg/ha) *and high macrophyte-infested beels* (792 ± 192 kg/ha). *The interaction effect was significant* F (2,18) = 23.42, p = 0.001.

7.5 Repeated Measures ANOVA (RMANOVA)

RMANOVA is an extension of paired t-test. It is a test that seems close to one-way ANOVA as it allows checking for differences between the means of three and more groups. The essential difference is that the groups are dependent or related. This means that the groups contain data or measurements from the same individuals. What differs in terms of design may be (1) when measurements were taken—e.g. measurements were thus repeated in time, before, during and after an experiment or every fortnightly during a given time frame etc.—or (2) in which circumstances or condition measurements were performed, e.g., measurements were done three times, each time following application of a different feed. For example, suppose we want to test the significant difference between the yield of beels after adoption of fish stock enhancement at three times (before, during and after an intervention). The null hypothesis states that the means of all groups are equal.

Assumptions

1. Each individual is represented in the form of a measurement in each of the tested groups (there cannot be any missing value).
2. Normality of distribution.
3. Sphericity, which means that equality of variance is verified when comparing any two groups (all possible pairs of groups) in the experimental design.
4. Groups do not contain outliers.

Statistical Analysis Data collected on repeated measures can be analysed like one-way ANOVA.

7.6 Analysis of Variance and Experimental Designs

The analysis of variance encompasses a broad spectrum of experimental designs ranging from the simple to the complex. A brief discussion on various designs of experiments is discussed in the following section.

7.7 Experimental Designs

One of the main objectives of designing an experiment is how to verify the hypothesis in an efficient and economical way. ANOVA technique, which is based on some assumptions, can be used to test null hypothesis of equality of several means of normal populations having same variances. If these assumptions are violated, the outcome of the test of hypothesis then may also be faulty that leads to erroneous conclusions. The mechanism to obtain the data such that the assumptions are met and the data is readily available for the application of tools like ANOVA is achieved by the design of experiment. Design of experiment provides a method by which the treatments are placed at random on the experimental units in such a way that the responses are estimated with the utmost precision possible. The basic terminologies used in the experimental design are as follows (see Table 7.8):

Experiment It is a method of getting an answer to a question which the experimenter wants to know.

Table 7.8 Different design of experiment and their analytical methods

ANOVA	Design of experiment
One-way ANOVA	Extension of unpaired t-test when we compare means of more than two groups. One factor with different levels. Completely randomized design (CRD)
One-way repeated measures ANOVA	Extension of paired *t*-test. It is used when the data are longitudinal and are measured on each individual at each successive time point. Completely randomized design (CRD)
Two-way ANOVA	Two factors with same or different levels. Randomized block design (RBD)
Three-way ANOVA	Two factors with same or different levels. Latin square design (LSD)
More complex ANOVA	Hierarchical (nested) designs, splitplot design, strip- plot design, etc.

Experimental Unit The experimental unit is the physical entity which can be assigned to, at random, a treatment. It is also the unit of statistical analysis. For example, the experimental unit may be an individual fish, the cage of fishes, tanks, aquaria, etc.

Treatment Various objects of comparison in an experiment are called treatments. In a cage culture experiment, fish weight is observed for different types of feeds. Here different feeds are being compared as treatments.

Sampling Unit The object measured in an experiment is called the sampling unit. It may be different from the experimental unit.

Factor A factor is a variable defining a categorization. It can be fixed (all the levels of interest are included in the experiment) or random (if all the levels of interest are not included in the experiment). For example, gender or sex is a factor variable with two levels, female and male.

Replication It is the repetition of the experimental situation by replicating the experimental unit.

Experimental Error The random part of variation that remains unexplained in any experiment is termed as experimental error. An estimate of experimental error can be obtained by replication.

Treatment Design A treatment design is the manner in which the levels of treatments are arranged in an experiment.

7.7.1 Principles of Experimental Design

There are three basic principles of design of experiment which were developed by Sir Ronald A. Fisher for the validity of statistical analysis and enhancing the precision of the experiments.

1. **Randomization:** Randomization is a process to allocate the treatments in the experimental units randomly to avoid any bias in the experiment which may be conscious or unconscious and ensures the validity of the results. It is a process in which every experimental unit has equal chance of receiving each treatment. It also ensures independence of the observations which is necessary for drawing valid inference from the observations by applying appropriate statistical techniques.
2. **Replication:** In the replication principle, any treatment is repeated a number of times to obtain a valid and more reliable estimate than which is possible with one observation only. It also provides an estimate of the error variance. Replication provides an efficient way of increasing the precision of an experiment. The precision increases with the increase in the number of observations. Replication provides more observations when the same treatment is used, so it increases precision. There should be much uniformity or homogeneity within each replication. A general rule is to get as many replications which will provide at least 12 degrees of freedom for the error. Usually one should not use less than four replications.
3. **Local control:** The replication is used with local control to reduce the experimental error. The process of reducing the experimental error by dividing the relativity heterogeneous experimental area into homogeneous blocks is known as local control and used to increase the efficiency of the design of experiment.

7.8 Completely Randomized Design (CRD)

The CRD is the simplest design. In this design, all experimental units are considered homogeneous, and no division or grouping among them exists. The treatments under study are allocated randomly to the whole set of experimental units without making any effort for more homogeneity. It is entirely flexible in the sense that any number of treatments or replications (equal or unequal) may be used. It is used under controlled condition and studies the effects of one factor, keeping other factors constant, known as single factor experiment. In the field of fisheries, suppose we want to test the different feeds' efficacy in relation to fish survival and growth in the identical and uniform fish (similar size and species) cages. The feed are of three types such as commercial feed, farm-made feed with soya protein content (10%) and farm-made feed without soya protein content (10%). We replicate these three feeds four times; then the total experimental unit will be 3×4 equals 12. So, here we need 12 uniform cages. The water factors (water level), especially environmental and water chemistry, e.g. temperature, DO, pH, alkalinity, ammonia levels, nitrite levels, etc., should be measured or recorded and made sure to be the same for all the tanks. Then equal number of fishes of same size (say, 30 no.) is placed in each of the water-filled cages, and feed are randomly placed in the tanks (using random number table or lottery system). Once randomization is complete, the researcher can start the experiment and collect required data depending on the specific objectives of the research. Usually here only one source of variation exists, that is, feed factor. The data generated from this design is analysed in the same way as that of one-way ANOVA.

7.9 Randomized Block Design (RBD)

The randomized complete block design (RCBD) is probably the most widely used design because, in reality, it is difficult to find all identical or uniform experimental units. In the field of fisheries science, experimental cages are set up in large water bodies like floodplain wetlands (beels) or reservoirs or tanks. There is a chance of low variability or higher chance for any treatment to show its real impacts. However, due care should be taken if cages are (1) at or near the bank of beels or reservoirs, where sunlight or shade can affect the rows at or near the bank more than others, and (2) close to the water source or human activities, such as boating. These are unavoidable factors which have large effects on the response variable. In this condition, their effects are removed by way of blocking in the experimental design. Blocking makes the experimental unit more homogeneous within the block. Each block is sub-divided into equal number of treatments called plots. Each of the treatment is placed in each of the experimental plot (tanks or cages) randomly for each block in such a way that every treatment is placed in each blocks. Let four treatments (feeds) have to be compared; each of the feed is replicated four times in four blocks, and then 4 × 4 × 4 equals to 64 number of experimental unit (tanks/ponds) are required. Here we have two sources of variation, one variation due to block and second variation due to treatments (feed). Hence, the analysis of the generated data will be performed by two-way ANOVA. The data generated from this design is analysed in the same way as that of two-way ANOVA. Diagrams of most common forms of experimental designs analysed by ANOVA are shown in Fig. 7.1.

One-way ANOVA

Diet group		
Control	Feed 1	Feed 2
x	x	x
x	x	x
x	x	x
x	x	x
x	x	x
x	x	
x		

Two-way ANOVA

	Macrophyte infestation		
Beel	Low	Moderate	High
Stocked	xxxx	xxxx	xxxx
	xxxx	xxxx	xxxx
	xxxx	xxxx	xxxx
	xxxx	xxxx	xxxx
Unstocked	xxxx	xxxx	xxxx
	xxxx	xxxx	xxxx
	xxxx	xxxx	xxxx
	xxxx	xxxx	xxxx

One-way repeated measures ANOVA

	Intervention		
Beel	Before	During	After
1	x	x	x
2	x	x	x
3	x	x	x
.	.	.	.
.	.	.	.
.	.	.	.
k	x	x	x

Fig. 7.1 Schematic data format for different ANOVA

7.10 Conclusion

This chapter describes some of the parametric test of significance based on ANOVA and design of experiments and their application in inland fisheries. Analysis procedures using SPSS are provided, which will be helpful to repeat the procedures for other data set. The knowledge gained from this chapter will help in designing experiment for data generation as well as in analysing experimental data in the future. For further study, the readers can access the list of suggested readings as listed at the end of the chapter.

Exercise

1. Analysis of variance is a statistical method of comparing the........... of several populations.
   (a) Standard deviations
   (b) Variances
   (c) Means
   (d) Proportions
2. The ....sum of squares measures the variability of the observed values around their respective treatment means.
   (a) Treatment
   (b) Error
   (c) Interaction
   (d) Total
3. The............sum of squares measures the variability of the sample treatment means around the overall mean.
   (a) Treatment
   (b) Error
   (c) Interaction
   (d) Total
4. If the true means of the k populations are equal, then MST/MSE should be:
   (a) More than 1.00
   (b) Close to 1.00
   (c) Close to 0.00
   (d) Close to −1.00
5. If the MSE of an ANOVA for six treatment groups is known, you can compute.
   (a) DF
   (b) The standard deviation of each treatment group
   (c) The pooled standard deviation
   (d) None of the above
6. To determine whether the test statistic of ANOVA is statistically significant, it can be compared to a critical value. What two pieces of information are needed to determine the critical value?

(a) Sample size, number of groups
(b) Mean, sample standard deviation
(c) Expected frequency, obtained frequency
(d) MSTR, MSE

7. Which of the following is an assumption of one-way ANOVA comparing samples from three or more experimental treatments?
   (a) All the response variables within the k populations follow a normal distribution.
   (b) The samples associated with each population are randomly selected and are independent from all other samples.
   (c) The response variable within each of the k populations has equal variances.
   (d) All of the above.
8. The error deviations within the SSE statistic measure distances:
   (a) Within groups
   (b) Between groups
   (c) Both (a) and (b)
   (d) None of the above
9. When the *k* population means are truly different from each other, it is likely that the average error deviation:
   (a) Is relatively large compared to the average treatment deviations
   (b) Is relatively small compared to the average treatment deviations
   (c) Is about equal to the average treatment deviation
   (d) None of the above
   (e) Differ significantly between at least two of the populations
10. As variability due to chance decreases, the value of F will
   (a) Increase
   (b) Stay the same
   (c) Decrease
   (d) Can't tell from the given information
11. In a study, subjects are randomly assigned to one of three groups: control, experimental A or experimental B. After treatment, the mean scores for the three groups are compared. The appropriate statistical test for comparing these means is:
   (a) The correlation coefficient
   (b) Chi square
   (c) The *t*-test
   (d) The analysis of variance
12. In one-way ANOVA, which of the following is used within the F-ratio as a measurement of the variance of individual observations?
   (a) SSTR
   (b) MSTR
   (c) SSE
   (d) MSE

13. When conducting an ANOVA, FDATA will always fall within what range?
    (a) Between negative infinity and infinity
    (b) Between 0 and 1
    (c) Between 0 and infinity
    (d) Between 1 and infinity
14. You obtained a significant test statistic when comparing three treatments in a one- way ANOVA. In words, how would you interpret the alternative hypothesis H_A?
    (a) All three treatments have different effects on the mean response.
    (b) Exactly two of the three treatments have the same effect on the mean response.
    (c) At least two treatments are different from each other in terms of their effect on the mean response.
    (d) All of the above
15. If the sample means for each of k treatment groups were identical (yes, this is extremely unlikely), what would be the observed value of the ANOVA test statistic?
    (a) 1.0
    (b) 0.0
    (c) A value between 0.0 and 1.0
    (d) Infinite
16. Suppose the critical region for a certain test of the null hypothesis is of the form $F > 9.48773$ and the computed value of F from the data is 1.86. Then:
    (a) H_0 should be rejected.
    (b) The significance level is given by the area to the left of 9.48773 under the appropriate F distribution.
    (c) The significance level is given by the area to the right of 9.48773 under the appropriate F distribution.
    (d) The hypothesis test is two-tailed
17. What would happen if instead of using an ANOVA to compare ten groups, you performed multiple *t*-tests?
    (a) Nothing, there is no difference between using an ANOVA and using a *t*-test.
    (b) Nothing serious, except that making multiple comparisons with a *t*-test requires more computation than doing a single ANOVA.
    (c) Sir Ronald Fischer would be turning over in his grave; he put all that work into developing ANOVA, and you use multiple *t*-tests
    (d) Making multiple comparisons with a *t*-test increases the probability of making a Type I error.
18. What is the function of a post-test in ANOVA?
    (a) Determine if any statistically significant group differences have occurred.
    (b) Describe those groups that have reliable differences between group means.
    (c) Set the critical value for the F test (or chi-square).
    (d) None of the above

19. In ANOVA with 4 groups and a total sample size of 44, the computed F-statistic is 2.33. In this case, the p-value is:
    (a) Exactly 0.05
    (b) Less than 0.05
    (c) Greater than 0.05
    (d) Cannot tell— it depends on what the SSE is
20. A fisheries researcher wishes to test the difference in mean weights of a single species of fish caught by fishermen in three different wetlands in Uttar Pradesh. What is the correct statistical analysis?
    (a) Two-way ANOVA
    (b) Repeated ANOVA.
    (c) One-way ANOVA
    (d) None of the above

Further Reading

Anderson TW, Sclove SL (1974) Introductory statistical analysis, Houghton Mifflin Company
Bhujel RC (2008) Statistics for aquaculture, 1st edn. Wiley
Biradar RS (2002) Fisheries statistics, 2nd edn. Central Institute of Fisheries Education, Mumbai
Clarke GM, Cooke D (1998) A basic course in statistics. Arnold, Electronic Statistics Textbook http://www.statsoftinc.com/textbook/stathome.html.
Roy AK, Sarangi N (2008) Applied bioinformatics, statistics and economics in fisheries research. New India Publishing Agency

Concept of Simple Correlation and Regression

8

Key Topics

- Concept of correlation
- Test of significance of correlation
- Non parametric correlation
- Concept of regression
- Properties of regression coefficient
- Testing of regression parameters
- Model accuracy and outlier detection
- Case study for fitting regression model in fisheries data

Acorrelation coefficient (r) is a statistical measure of linear association between or among variables. Correlation can be either positive or negative and approach to zero that depends on the situation and the variables under study. When an increase in one variable is paired with the increase in the other variable, then it is known as positive correlation. Length-weight relationship of fish is a popular example. However in the case of negative correlation, increased value of the one variable is coupled with decreased value of the second variable. For example, fish survival and daily weight gain are normally observed as negative correlation because high survival means more fish remained in the system and share limited space and food, which results in smaller fish, i.e. low daily weight gain.

B. K. Das et al., *Concept Building in Fisheries Data Analysis*,
https://doi.org/10.1007/978-981-19-4411-6_8

8.1 Simple Correlation Coefficient

Karl Pearson, a British biometrician, developed a formula to compute correlation. Correlation coefficient between two random variables X and Y, usually denoted by $r(X, Y)$ or simply r_{xy}, is a numerical measure of linear relationship between them and defined by

$$r(X, Y) = \frac{\text{Covariance}}{\sigma_x \sigma_y};$$

$$r(X, Y) = \frac{\sum (X - \overline{X})(Y - \overline{Y})}{\sqrt{\left[n \sum X^2 - (\sum X)^2\right]\left[n \sum Y^2 - (\sum Y)^2\right]}}$$

The formula can be rewritten without measuring the means, and the resulting correlation coefficient is also formulated as

$$r(X, Y) = \frac{\sum XY - (\sum X)(\sum Y)}{\sqrt{\left[n \sum X^2 - (\sum X)^2\right]\left[N \sum Y^2 - (\sum Y)^2\right]}}$$

where N is the number of data pairs; ΣXY is the sum of the products of data pairs; ΣX is the sum of the first variable, i.e. X; ΣY is the sum of the second variable, i.e. Y; ΣX^2 is the sum of the squared observations of variable X; ΣY^2 is the sum of the squared observations of variable Y; σ_x = standard deviation of x; and σ_y = standard deviation of y.

8.1.1 Some Properties of Correlation Coefficient

Karl Pearson's correlation coefficient is also called as product-moment correlation coefficient. The correlation coefficient is measured within ± 1 scale or $-1 \leq r \leq +1$. If two variables have a correlation value of +1, that means they have perfect (100%) positive linear correlation (see Fig. 8.1); similarly, if they have a correlation value of −1, that means perfect (100%) negative linear correlation (see Fig. 8.1). Similarly, correlation of zero (0) means the two variables are not linearly correlated.

However, a correlation coefficient of exactly zero does not occur in reality. Sometimes correlation coefficient is found to be zero between two variables X and Y. Then we cannot conclude that the relationship is zero. It only means that there is no linear relationship between X and Y, but there is a chance of nonlinear relation between X and Y (Figs. 8.2, 8.3, and 8.4).

There is a strong quadratic relationship between X and Y in the right side of the scatter plot and almost no relationship in the left-hand side plot. But the correlation coefficient between x and y in both cases is 0.02 which is almost zero (see Fig. 8.5).

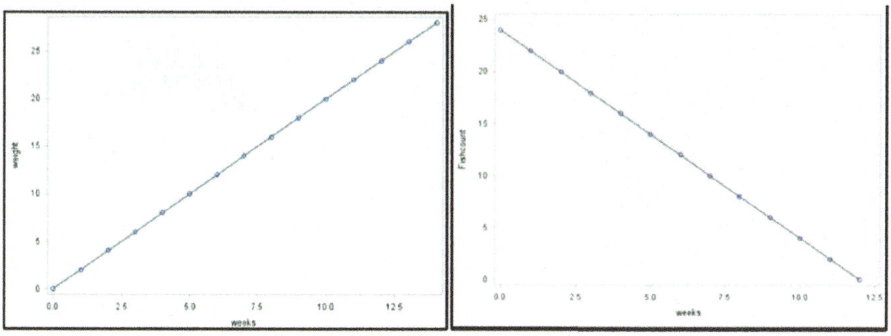

Fig. 8.1 $r = 1$, perfect positive correlation, and $r = -1$, perfect negative correlation

Fig. 8.2 Strong correlation: (**a**) positive and (**b**) negative

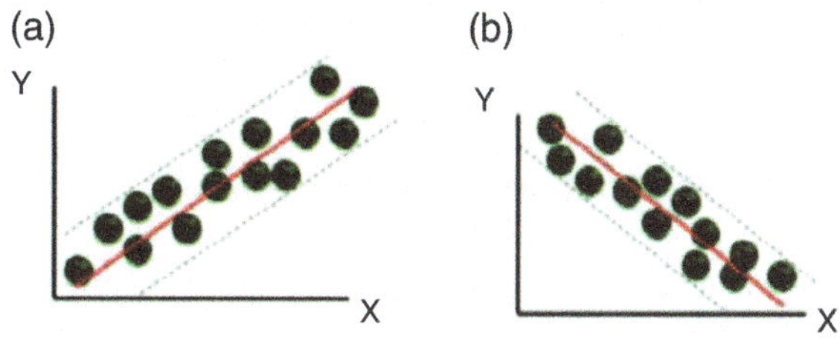

Fig. 8.3 Weak correlation, (**a**) positive and (**b**) negative

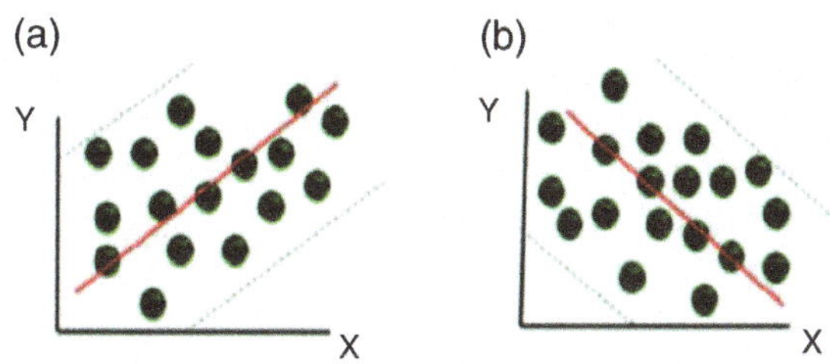

Fig. 8.4 Almost zero correlation

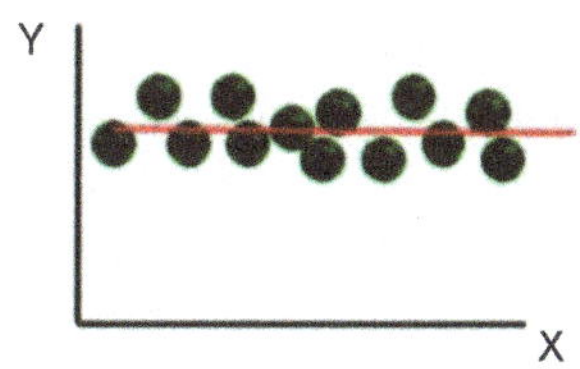

Note *Correlation is independent of change of origin and scale* (i.e., *if a constant is added, deleted, multiplied and divided to the variables, then their correlation will not be changed).*

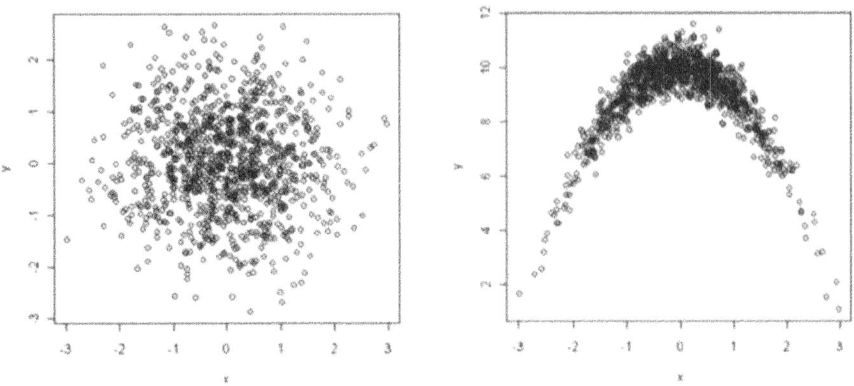

Fig. 8.5 Almost zero correlation with no linear relationship but strong nonlinear relationship

8.1.2 Distinction Between *r* and r^2

The squared correlation coefficient, i.e. r^2, represents the proportion of common variation between two variables in the given data sets. It is also called the coefficient of determination and expressed in percentage, e.g. 98%, which means 98% of the variation is due to their association and the remaining 2% is due to other factors or conditions, also called residuals. In general to understand the correlation value in qualitative sense, correlation (r) has been categorized as high (0.8–1.0), medium (0.6–0.8), fair (0.4–0.6) and low (<0.4). It is always recommended to present the results in terms of r^2 instead of r.

8.2 Test of Significance of Correlation Coefficient

However, the significance of correlation has to be tested for its reliability, which depends on some assumptions. The basic assumption is that the paired samples are a representative of a bivariate normal population. Sample size more than 100 is always assumed to be fairly large, and a researcher need not to worry about normality assumptions. Correlation coefficient is usually tested as to whether they are significantly different from zero (0) or not. The significance of correlation coefficients depends on their magnitudes as well as the size of the samples. Although a low correlation coefficient from a large sample size can be significant, reliability of the significance of the correlation coefficient increases with its magnitude. A difference of 0.10 between two correlation coefficients may not be significant if the two coefficients are 0.10 and 0.20, but the same degree of difference, i.e. 0.10, can be highly significant when the two coefficients are 0.80 and 0.90 in the same data sets. For the purpose of describing the method, an example is given.

Table 8.1 Length and weight of the *S. strongylura* collected from a lake and their square value (L^2 & Wt^2) and product value ($L \times Wt$)

L (mm)	Wt (gm)	$L \times Wt$	L^2	Wt^2
294	28	8232	86,436	67,765,824
229	12	2748	52,441	7,551,504
200	9	1800	40,000	3,240,000
315	35	11,025	99,225	121,550,625
265	22	5830	70,225	33,988,900
234	15	3510	54,756	12,320,100
252	17	4284	63,504	18,352,656
240	15	3600	57,600	12,960,000
218	11	2398	47,524	5,750,404
252	19	4788	63,504	22,924,944
259	20	5180	67,081	26,832,400
358	71	25,418	128,164	646,074,724
200	11	2200	40,000	4,840,000
223	15	3345	49,729	11,189,025
178	7	1246	31,684	1,552,516
188	9	1692	35,344	2,862,864
194	8	1552	37,636	2,408,704

Example 8.1
From a trial of length (L) and weight (W), measurements of 17 *S. strongylura* were recorded from the lake simultaneously (see Table 8.1)) to see the correlation. For this, the additional calculation which is required is the cross products of L and W, and their squares need to be calculated as shown in Table 8.1. Let us assume $\Sigma L = 4099$; $\Sigma W = 324$; $\Sigma LW = 88{,}848$; $\Sigma L^2 = 1{,}024{,}853$; and $\Sigma W^2 = 1{,}002{,}165{,}190$, and we found $r = 0.91$ and $r^2 = 0.93$. This shows that length and weight of *S. strongylura* have high correlation (0.934). In other words, it can be concluded that there is 93% variability of weight is explained by length.

The estimated standard error of 'r' is given by

$$\widehat{SE}(r) = \sqrt{\frac{1 - r^2}{n - 2}},$$

here r = correlation coefficient and n = number of pairs of observation.

One Sample Case
X and Y from which n pairs of samples have been drawn. The population correlation is generally denoted by ρ. Now test that the population correlation coefficient ρ from which the present sample is drawn is either 0 or different from 0, i.e. H_0: $\rho = 0$ against alternative hypothesis H_1: $\rho \neq 0$. The test statistics is given by

$$t = \frac{r\sqrt{n-2}}{\sqrt{1-r^2}} \tag{8.1}$$

If the computed value of $|t|$ exceeds the table value of t with (n–2) degrees of freedom (d.f.) at a given level of significance (say 5% or 1%), then the null hypothesis is rejected, and it may be concluded that the variables are correlated in the population.

It may be of interest to test the population correlation coefficient has some specified values other than zero. The test hypothesis in such case is, i.e. H_0: $\rho = \rho_0$ against alternative hypothesis H_1: $\rho \neq \rho_0$. In this situation r and are transformed as follows:

$$Z = \frac{1}{2}\log_e\left(\frac{1+r}{1-r}\right)$$
$$Z_0 = \frac{1}{2}\log_e\left(\frac{1+\rho_0}{1-\rho_0}\right)$$

The statistics U is computed as $U = (Z - Z_0)\sqrt{n-3}$ which follows standard normal distribution. If the absolute value of $|U| > 1.96$, then we reject the null hypothesis, H_0: $\rho = \rho_0$ at 5% level of significance.

Two Sample Case

Let r_1 and r_2 be the sample correlation coefficients between two variables obtained from two independent samples of sizes n_1 and n_2, respectively. Here the interest is to test the two samples drawn are from the same populations with the same correlation coefficient ρ (say), i.e. H_0: $\rho = r_2 =$ against H_0: $r_1 \neq r_2$. To test this hypothesis, the test statistics is given by where $Z = \frac{1}{2}\log_e\left(\frac{1+r_i}{1-r_i}\right); i = 1, 2$ and $S.E.(Z_1 - Z_2) = \sqrt{\frac{1}{(n_1-3)} + \frac{1}{(n_2-3)}}$.

Under the null hypothesis, Z is $N(0,1)$, so we apply the standard normal deviate test under which the hypothesis is rejected at 5% level, if $Z > 1.96$, and 1% level of significance if $Z > 2.58$.

8.3 Non-parametric Correlation or Rank Correlation

When the character under study is not measured quantitatively (may be qualitative), then it can be ranked according to some criteria, and then its correlation is tested by non-parametric method such as Spearman's rank correlation. Suppose a researcher wants to test the correlation between egg viability (%) and the ammonia nitrogen level of the water in the breeding ponds—the relevant data is presented in Table 8.2.[1] In this situation both the variables are expressed as ranks, and rank correlation

[1]Source: Bhujel (2008).

Table 8.2 Egg viability and ammonia nitrogen level

Weeks	Viability (%)	Viability (ranks)	NH_3-N (PPM)	NH_3 (ranks)	d_i	d_i^2
1	74.9	1	38	12	−11	121
2	77.6	2	38.8	11	−9	81
3	80.02	4	37.1	9	−5	25
4	81.7	6	38.2	10	−4	16
5	78.2	3	34.5	8	−5	25
6	85.6	10	33.2	7	3	9
7	81.1	5	32.1	5	0	0
8	85.5	9	31.9	3.5	5.5	30.25
9	83	7	32.4	6	1	1
10	86.4	11	31.9	3.5	7.5	56.25
11	85.3	8	30.5	2	6	36
12	87.2	12	29.3	1	11	121

coefficient is calculated. The correlation between the ranks of two variables is called the rank correlation. Let (x_i, y_i); i = 1, 2. . . n be the ranks of two characters; let us say fish egg viability (%) and ammonium nitrogen level, respectively.

The Spearman's correlation coefficient is calculated as

$$\rho(X, Y) = 1 - \frac{6\sum_{i=1}^{n} d_i^2}{n(n^2 - 1)}$$

where $d_i = x_i - y_i$; $i = 1,2,\ldots, n$ denotes the difference between the ranks of ith individual. The limit for rank correlation coefficient is $-1 \leq \rho \leq 1$. Here $n = 12$, $\Sigma d^2 = 521.5$, and $\rho = -0.98$ (calculated value).

Environmental variables are associated with each other by a number of factors. For a researcher it is very important to select variables for data collection, and it is basically guided by the objectives of the research. Some of the variables are intermediate in nature which might be affected by other variables or factors. In this situation one should select the actual variables or factors. Length and weight of fish can vary simultaneously, caused by the amount or rate of feeding. In this case, it is clear that length and weight of fish are variables that have high correlation and the rate or amount of feeding is a factor. However, fish weight and length may show positive correlation with intermediate variables, such as DO levels and plankton growth, and negative correlation with ammonia and nitrite levels.

Note *Sample correlation coefficient is the estimate of population correlation, so if population data are not available, then sample correlation coefficient r is the approximated value of the population correlation.*

The following SAS code provides the correlation coefficient and their significance level.

SAS Code for Karl Pearson's Correlation Coefficient Analysis

```
data fish;
input length weight; cards;
cards;
294 28
229 12
....
.... ;
proc corr data =fish;
run;
```

8.4 Concept of Regression Analysis

In the fisheries field, a researcher is very much interested in the cause and effect relationship and its significance. The factor which is also known as cause brings a significant or insignificant change (effect) on certain characteristics (response) which is simply analysed by the regression analysis. Regression analysis is a mathematical measure of the average relationship between two or more variables in terms of the original data. So, it is a statistical tool for investigating the relationship between a dependent variable and one or more independent (explanatory) variables. For example, the increase in nitrogen fertilizer level in the pond increases the fish yield, and increase in feeding rate increases the growth rate, etc. While designing survey or experimental research, it becomes very important to identify the factor and the response variable. The level of factor should be appropriate (low or high) to establish the complete relationship. In general, three levels or more are found to be good, but sometimes researcher uses less level due to insufficient experimental unit. There is always a confusion regarding the replication, which should be at least three also. For regression analysis, it is better to increase the level of the factor rather than have the replication for the levels. For example, if a researcher has six ponds for a trial, it is better to have six treatment levels, e.g. 0, 1, 2, 3, 4 and 5 kg N/ha/day, than to have three treatment levels, such as 0, 3 and 6 kg N/ha/day, with two replications (Bhujel, 2008). In this section simple, linear regression analysis is explained through exploring relationship in which only one factor is considered to affect the response variable assuming the other potential factors are constant.

Before going directly for regression analysis, we have to visualize the nature of the data carefully and then decide the appropriate analysis. Commonly by plotting the data distribution, one can identify the relationship as linear or nonlinear. If a factor increases or decreases in a constant manner, then the relationship is assumed to be linear. For example, in Fig. 8.6, one can guess that there is a linear relationship between fish production per year and nitrogen level. As the nitrogen level kg per hectare per day increases, the fish production (tone) per hectare per year increases.

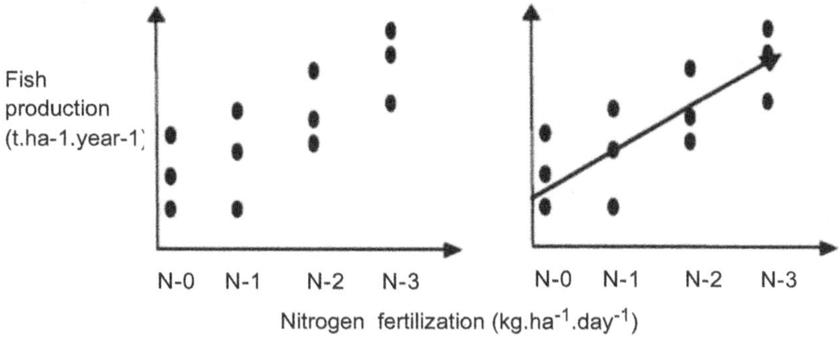

Fig. 8.6 The plots showed the fish production per year increases as the nitrogen level increases. (**Source:** Bhujel 2008)

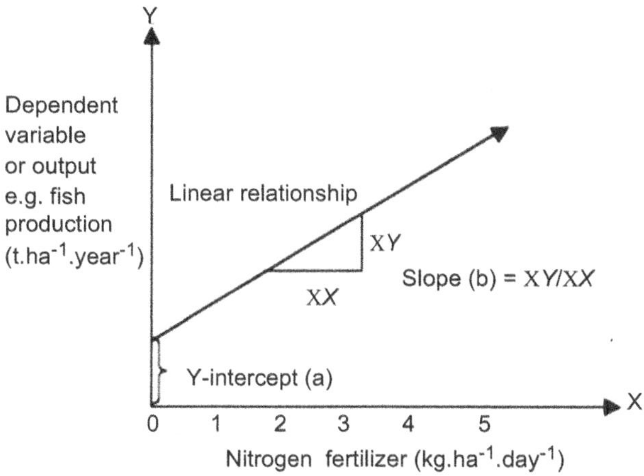

Fig. 8.7 Slope of the line as relationship between factor X (nitrogen fertilizer level) and response variable (fish production). (**Source:** Bhujel, 2008)

Once we identify the linear relationship, then linear regression analysis should be performed on the studied data sets. The main aim in linear regression is to find the trend line that the sample data represents.

The best fitted line is selected in such a way that it passes through the maximum number of points, and the sum of squared deviations to the points from the line is minimal. The method used for best fitted line is known as least square deviation (LSD) method. The next target after identifying the straight line is to find its slope (b) or gradient which can be explained as the change in response variable ($\triangle Y$) per unit change in factor ($\triangle X$), i.e. $b = \triangle Y/\triangle X$ (Fig. 8.7). In Fig. 8.7 our response variable is fish production (tone) per hectare per year, and the factor is the nitrogen fertilizer level. We observe three types of slope (b) or trend, i.e. positive, negative and zero.

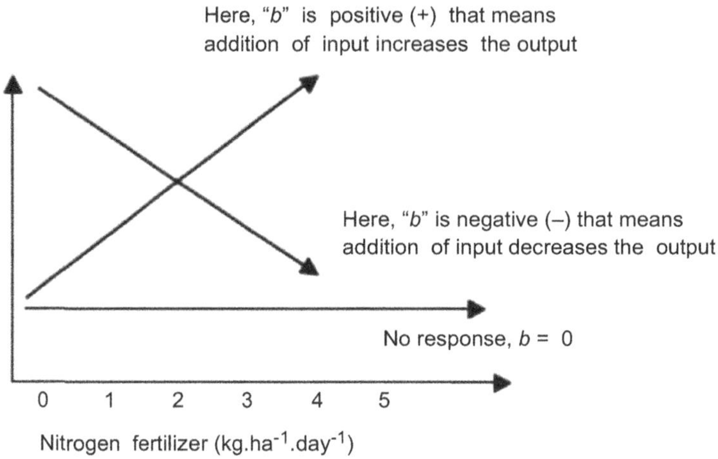

Fig. 8.8 Relationship between factor and variable which may be either positive (+), negative (−) or (0) zero depending on the situation. (**Source:** Bhujel 2008)

- Positive slope means there is an increase in response variable with the increase in factor.
- Negative slope means the response variable declines when the factor increases.
- Zero slope means there is no effect of factor levels on the response variable (Fig. 8.8).

The simple linear relationship is expressed mathematically as shown below:

$$Y = a + bX + e \tag{8.2}$$

where *Y* is a response variable that depends on factor *X*. *X* is a fixed factor that affects variable *Y*. *a* is the constant or intercept, i.e. the level of *Y* at 0 level of *X*. *e* is random error component (independent) with $e \sim N(0, \sigma^2)$. *b is* the slope of line or change in *Y* per unit change in factor *X*, i.e. $b = \Delta Y/\Delta X$; it is known as regression of *Y* on *X*; *b* can be estimated as b_{yx} (estimated value) and is also known as the unbiased estimator of *b*. ***n*** = number of paired (X_i, Y_i) observations

$$b_{yx} = \hat{b} = \frac{\sum\left[\left(X_i - \overline{X}\right)\left(Y_i - \overline{Y}\right)\right]}{\sum\left(X_i - \overline{X}\right)^2} = r_{xy}\frac{\sigma_y}{\sigma_x} \tag{8.3}$$

r_{xy} = correlation between *X* and *Y* and σ_y and σ_x are the standard deviation of the variable *Y* and *X*, respectively, and parameter *a* is estimated as $a = \overline{y} - b_{yx}\overline{x}$; where $\overline{y} = \frac{1}{n}\sum_{i=1}^{n} y_i$ and $\overline{x} = \frac{1}{n}\sum_{i=1}^{n} x_i$.

Thus there are always two lines of regression *Y* on *X* and *X* on *Y*. The line of regression of *Y* on *X* is used to estimate or predict the value of *Y* for any given value of *X*, i.e. *Y* is a dependent variable and *X* is explanatory variable and the estimate of the parameters obtained using the principle of least square method.

8.4.1 Properties of Regression Coefficient

1. Correlation coefficient is the geometric mean of the two mutual regression coefficients of two variables X and Y.
2. If one regression coefficient is greater than unity, then the other must be less than unity.
3. Arithmetic mean of the regression coefficients is greater than the correlation coefficient r, provided $r > 0$.
4. Regression coefficient is independent of change of origin but not of scale.
5. Regression coefficient range is $-\infty$ to $+\infty$.
6. The sign of correlation and regression is always the same.
7. If the variables X and Y are independent, then the regression coefficient is zero.

8.4.2 Testing of Regression Parameters

After identifying the slope of the regression analysis, next task is to test whether regression model is significant or not. The following terms should be understand for understanding the testing of regression model and parameters.

Sum of square due to regression (SSR) $= \sum_{i=1}^{n} (\widehat{y}_i - \overline{y})^2$

Sum of square due to errors (residuals) (SSE) $= \sum_{i=1}^{n} (y_i - \widehat{y}_i)^2$

Total sum of square (SST) $= \sum_{i=1}^{n} (y_i - \overline{y})^2$

Sum of square (SSx) $= \sum_{i=1}^{n} (x_i - \overline{x})^2$

Model variance is estimated by $= \widehat{\sigma^2} = \frac{\text{SEE}}{n-2}$.

$\widehat{y}$ = estimated response. The regression model significance is tested through ANOVA by F-statistics:

$$F = \frac{\text{SSR}/p}{\frac{\text{SEE}}{n-2}} \sim F_{\alpha,p,n-p-1};$$

if the calculated value is greater than the table value of given F, then we say model is significant at significance level α. Here p is the number of parameters to be estimated, and the total model parameters will be $(p + 1)$.

Note For single factor and response variable, $p = 1$.

The estimated parameter b is tested for linear relationship as follows:

Hypothesis

Null hypothesis H_0: $b = 0$ (no linear relationship)

Alternate hypothesis H_1: $b \neq 0$ (linear relationship)

Thus, the test statistics $= \frac{\widehat{b}-b}{\sqrt{\frac{\sigma^2}{S_{xx}}}} \sim N(0, 1)$, where $\widehat{b} = \frac{\sum (x_i-\overline{x})y_i}{\sum (x_i-\overline{x})^2}$.

Under null hypothesis H_0: $b = 0$, test statistics $Z = \frac{\widehat{b}}{\sqrt{\frac{\sigma^2}{S_{xx}}}}$.

Now, two condition arises: (a) σ^2 is known and (b) σ^2 is unknown.
For known, reject the null hypothesis H_0 if $|Z| > Z_{\frac{\alpha}{2}}$.

Foe unknown ρ^2, test statistics $t = \frac{\widehat{b}}{\sqrt{\frac{\mathrm{MS(SSE)}}{S_{xx}}}}$.

where, $\mathrm{MS(SSE)} = \frac{\mathrm{SSE}}{n-2}$.
Reject the null hypothesis H_0 if $| t |> t_{\frac{\alpha}{2}}, n-2$.

8.4.3 Model Accuracy Measurement and Outlier Detection

Further we test the model performance using coefficient of determination (variance explained by the model of the total variance) defined by $R^2 = \frac{\mathrm{SSR}}{\mathrm{SST}} = 1 - \frac{\mathrm{SSE}}{\mathrm{SST}}$ and the range of the R^2 is $0 \le R^2 \le 1$; if SSE = 0 or $R^2 = 1$, then we can say that the fitted model explains all the variation in the given data. Sometimes R^2 is not alone considered as the best model fitting criteria, then adjusted R^2 is used as statistics for this purpose, and then it is given by

$$R^2_{\mathrm{adj}} = 1 - \frac{n-1}{n-2}\left(1 - R^2\right)$$

And finally the adequacy of the model was tested by plotting the residual against the predicted value for the stochastic behaviour. And the outlier in the data sets is identified by the studentized residual plot. The Cook's D statistics and plot is used as a measure of the influence of the observation on the estimated regression coefficients. The SAS software is very useful for the linear regression analysis of the data. It gives the best linear fit line, model significance, regression parameter significance and model diagnostics statistics. The above example is analysed by SAS software and the output is given.

Example 8.2
A simple data set is given in Table 8.3 in which weights of fish (gm) are taken as response variable and the age of fish is taken as in weeks (factor) from 0 to 12th week which were found in a cage of reservoir.

Table 8.3 Weights of fish (gm) from 0 to 12th week of age

X; age (week)	*Y*; weight (g)	*X*; age (week)	*Y*; weight (g)
0	13.3	7	51.6
1	9.6	8	58.6
2	25.4	9	95.6
3	30.4	10	69
4	36.6	11	73.5
5	41.7	12	78.3
6	46		

Here:

Age of fish (weeks) is an independent factor X.

Weight of fish (g) is a dependent factor Y.

Total age (ΣX) = 78, $n = 13$, and mean age $(\overline{X}) = 6$ weeks.

Total weight (ΣX) = 639.6, $n = 13$, and mean weights $(\overline{Y}) = 49.2$ gm.

The linear model $Y = a + b$ X is fitted, and the regression coefficient (b) and its significance are computed as follows:

Sum of square due to regression (SSR) = $\sum_{i=1}^{n}(\hat{y}_i - \overline{y})^2 = 64449.20$

Sum of square due to errors (residuals) (SSE) = $\sum_{i=1}^{n}(y_i - \hat{y}_i)^2 = 937.31$

Total sum of square (SST) = $\sum_{i=1}^{n}(y_i - \overline{y})^2 = 7386.52$

Sum of square of X (Sxx) = $\sum_{i=1}^{n}(x_i - \overline{x})^2 = 182.00$

$$b_{yx} = \sum[(X_i - \overline{X})(Y_i - \overline{Y})] / \sum(X_i - X)^2$$

so, $b_{yx} = 5.95$, and a $= \overline{y} - b_{yx}\overline{x} = 13.5$, and R-square value can be computed as

$$R^2 = \left[\sum[(X_i - \overline{X})(Y_i - \overline{Y})]\right] / \sqrt{\sum(X_i - \overline{X})^2 \sum(Y_i - \overline{Y})^2}^2$$

$R^2 = 0.87$ and adjusted $R^2 = 0.86$

- The analysis can also be done using the SAS software. The results are shown as follows:
- SAS output of the regression model of fish weight gain as the response variable and age as factors is shown in Table 8.4.

Note *One observation is* ninth *is observed as outlier identified from the bivariate scatter plot and model fit diagnostic plot.* Fig. 8.9 *shows that there is one outlier and a linear upward trend exists in the data set.*

Table 8.4 Model parameter estimate and their significance and fit statistics using linear regression model fitted

Source	**DF**	**Sum of squares**	**Mean square**	***F* value**	**Pr > *F***
Model	1	6449.20	6449.20	75.69	<0.0001
Error	11	937.31	85.21		
Corrected total $R^2 = 0.8731$ Adj $R^2 = 0.8616$	12	7386.52			
Parameter estimates and their statistical significance					
Variable	**DF**	**Parameter estimate**	**Standard error**	***t* value**	**Pr > |*t*|**
Intercept	1	13.48352	4.83833	2.79	0.0177
Weeks	1	5.95275	0.68424	8.70	<0.0001

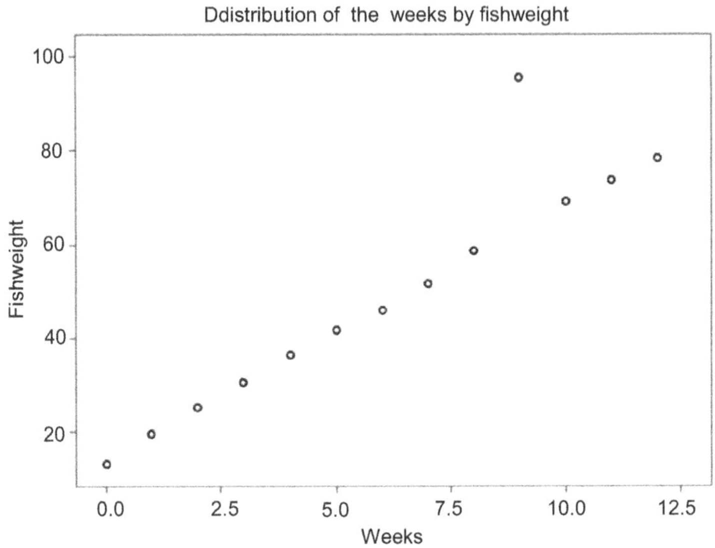

Fig. 8.9 The scatter plot of fish weight and age (weeks)

Table 8.4 showed the model and model parameters; both are significant at 5% level, and R square is 87%. Fig. 8.10 shows that residual follows the stochastic behaviour after fitting the model. Fig. 8.11 shows the line is fitted good and it lies between the range except one outlier point.

Note Fig. 8.12 *shows several regression diagnostic plots based on the residuals. The Rstudent and leverage plot showed that one outlier is present and it is identified by Cook's D plot.*

Example 8.3: Illustration of Inadequate Fitting
A case study is shown through an example of inadequate fitting of the linear regression model when response and factor relationship is not linear in nature. For this purpose a non-linear data is simulated using the help of Table 8.3.

The new data is generated by new weight as multiplying the weeks and square of the original weight from the above example data set in Table 8.3, and the new data set is presented in Table 8.5.

Linear regression model fitted on the new weight on age (weeks) and the SAS output is presented as follows. Fig. 8.13 shows that there is a quadratic relationship between the paired data sets with one outlier point, so by ignoring this information, we fit the linear regression analysis. Although the model and its parameters are also found to be significant and R^2 is 77% (see Table 8.8), fit diagnostic and residual plot

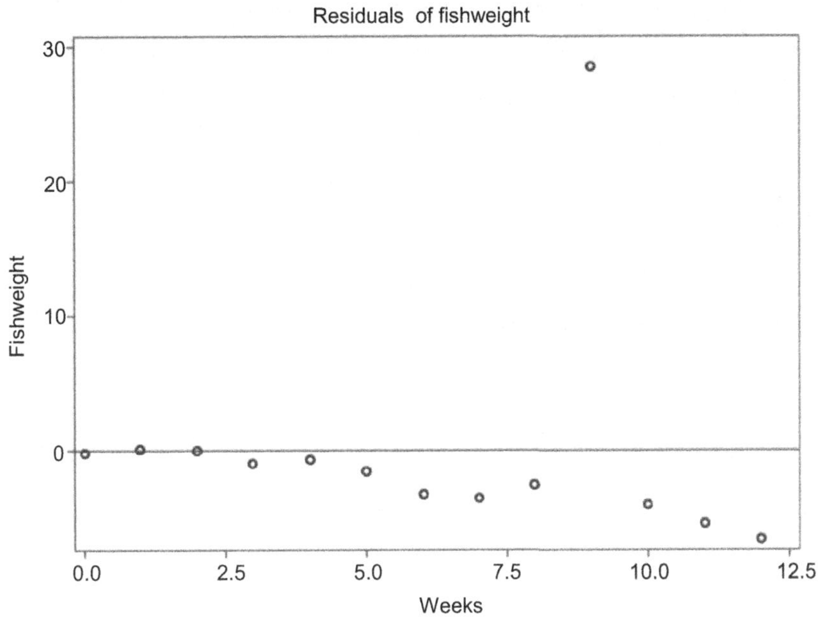

Fig. 8.10 Residual of weight vs weeks plot after fitting the model

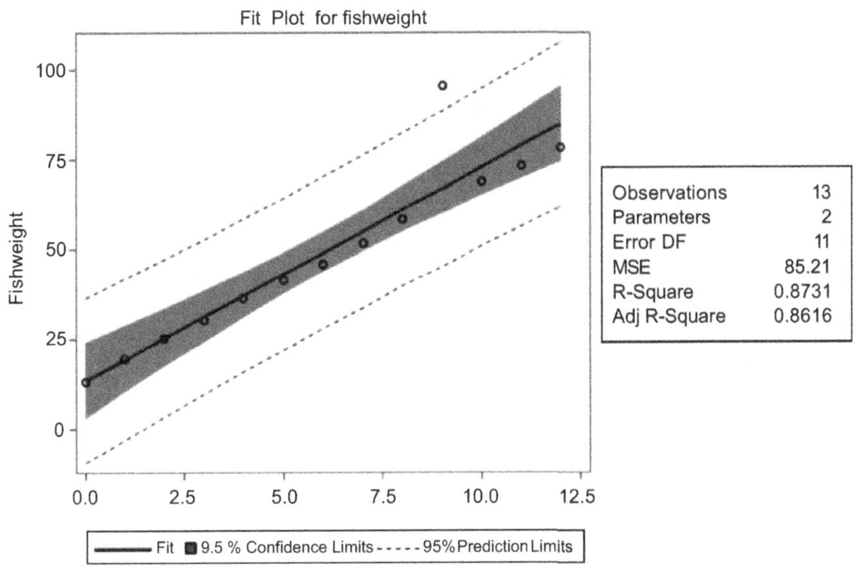

Fig. 8.11 Fit plot of fish weight against weeks

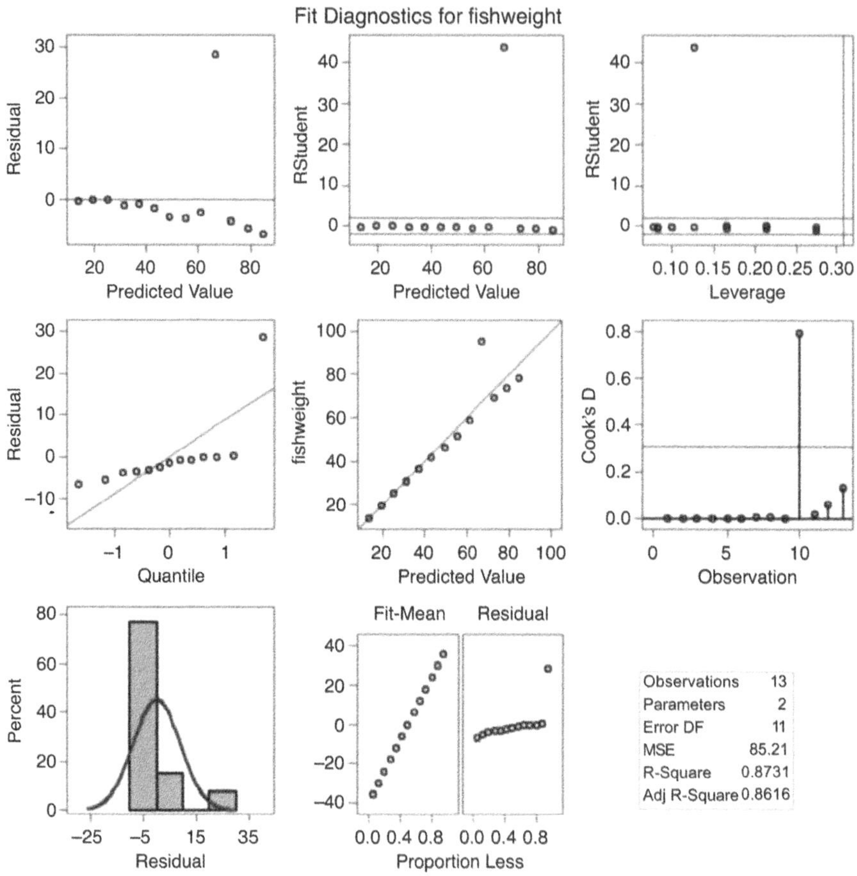

Fig. 8.12 Regression model fit diagnostic plot for fish weight gain at different age (weeks)

Table 8.5 New weight gain of the fish at different age (weeks)

Age (week)	Weight (gm)	Age (week)	Weight (gm)
0	0.00	7	18637.00
1	384.16	8	27471.68
2	384.16	9	82254.24
3	2772.48	10	47610.00
4	5358.24	11	59424.75
5	8694.45	12	73570.68
6	12696.00		

(quadratic behaviour) (see Figs. 8.14 and 8.15) showed that it does not follow the error assumptions. So, one should not fit the linear model when it is not following the linear relationship. Fit line (Fig. 8.16) showed that the observation and fit line lies outside the confidence interval of the model (Tables 8.6 and 8.7).

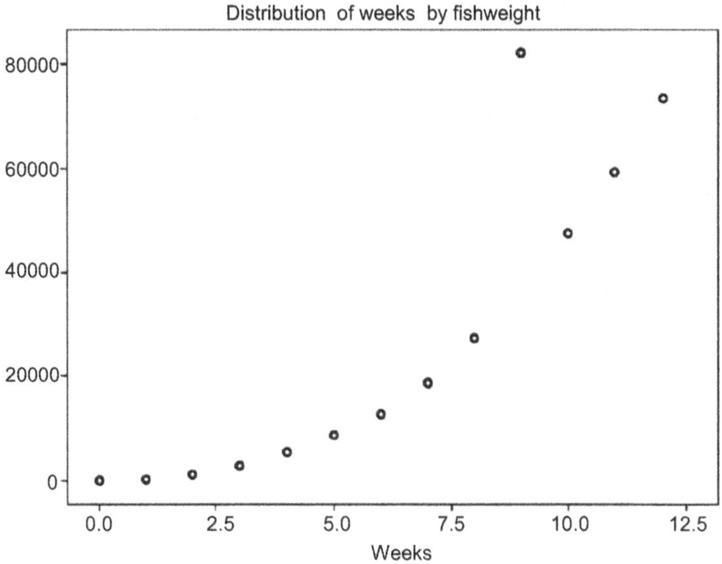

Fig. 8.13 Plot of fish weight versus age

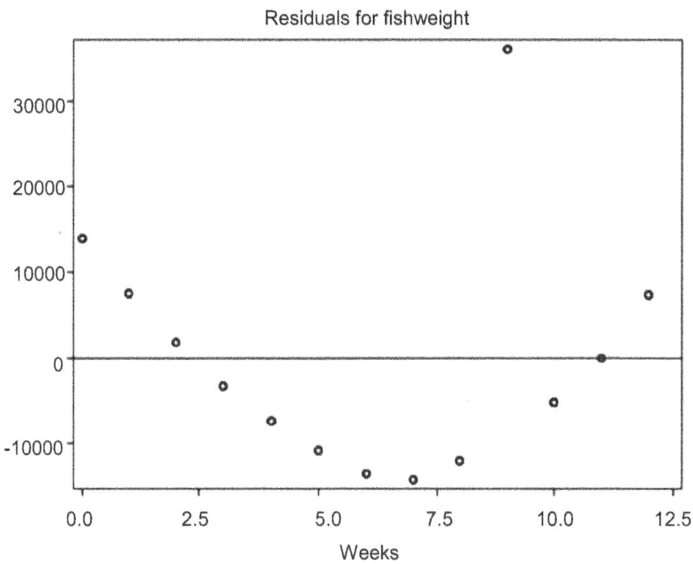

Fig. 8.14 Residual for fish weight plot vs weeks after fitting linear model

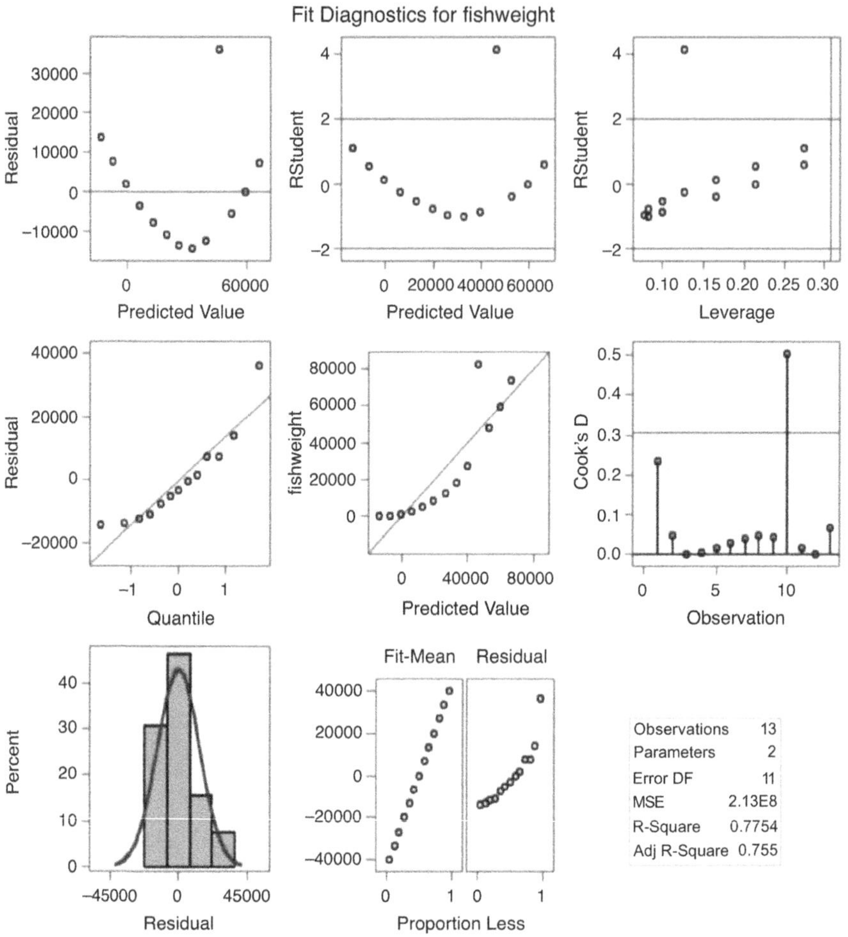

Fig. 8.15 Fit diagnostic plot after fitting linear model

Example 8.4: Illustration of Outlier Adjustment

For case study, we can remove or change the outlier point in Example 8.2, letting us change the ninth observation from 95.6 to 65.6 and then again fit it to the linear regression model and see the result. Results showed that if we compare Tables 8.4 and 8.5 fit statistics, the R square value increases from 87% to 99% and other parameters also changed. Residual plot (see Fig. 8.17) showed the stochastic behaviour, and Fig. 8.18 shows the best fit line (Tables 8.8 and 8.9).

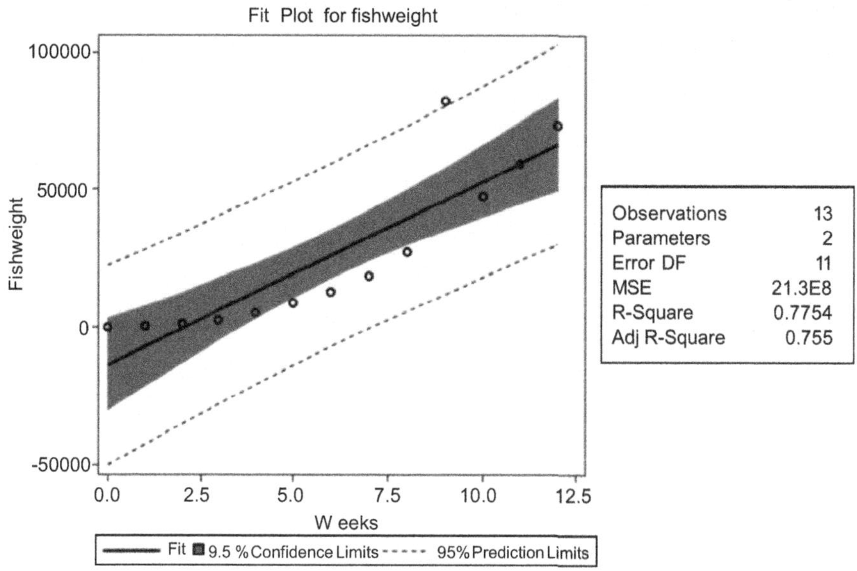

Fig. 8.16 Regression fit plot after linear model fitting on fish weight on age

Table 8.6 Model parameters estimate and their significance and fit statistics using linear regression model fitted

Source	DF	Sum of squares	Mean square	*F* value	Pr > *F*
Model	1	8,104,737,266	8,104,737,266	37.97	<0.0001
Error	11	2,347,801,357	213,436,487		
Corrected total $R^2 = 0.7754$ Adj $R^2 = 0.7550$	12	710,452,538,624			

Table 8.7 Parameter estimates and their statistical significance

Variable	DF	Parameter estimate	Standard error	*t* value	Pr > |*t*|
Intercept	1	−13,873	76571082.92564	−1.81	0.0974
Weeks	1	6673.19440	1082.92564	6.16	<0.0001

Note *There is always a chance of improving in model and its parameters by fitting, taking into account the assumptions and fit criteria of the model. One should be very careful to fit the best model for the given data sets.*

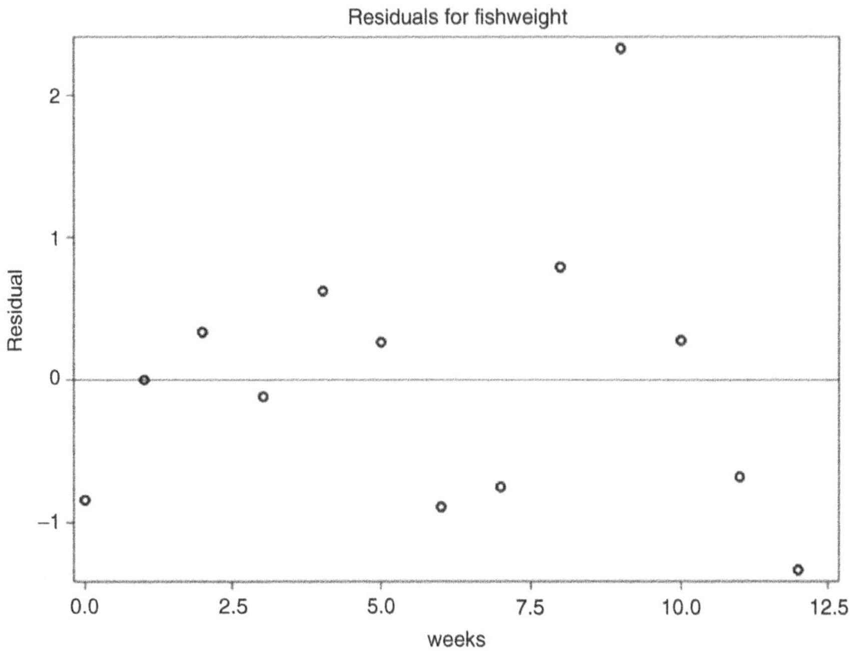

Fig. 8.17 Residual for fish weight plot vs weeks after fitting linear model

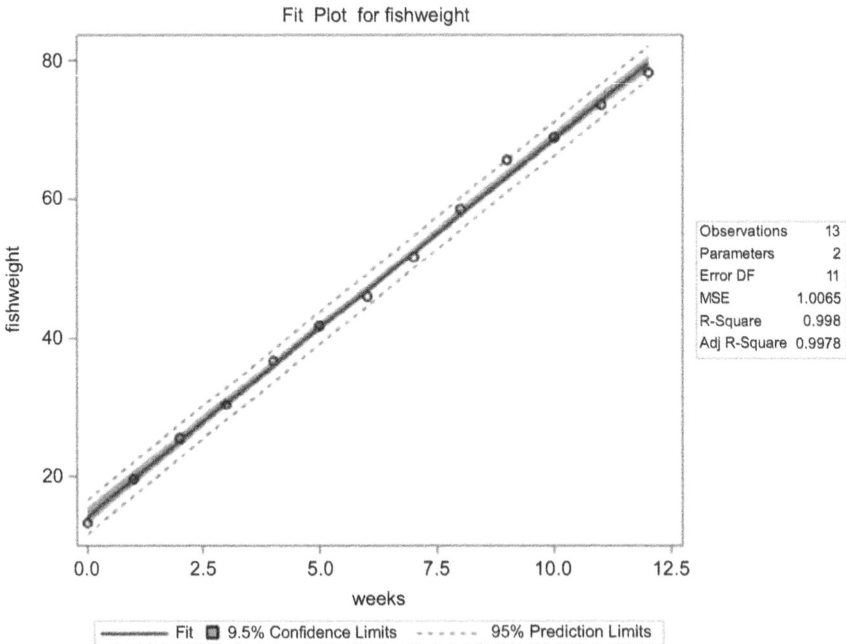

Fig. 8.18 Best fit line after changing the outlier and fit the regression line

Table 8.8 Model parameters estimate and their significance and fit statistics using linear regression model fitted

Source	DF	Sum of squares	Mean square	F value	Pr > F
Model	1	5422.21736	5422.21736	5387.02	<0.0001
Error	11	11.07187	1.00653		
Corrected total $R^2 = 0.9980$, Adj $R^2 = 0.9978$	12	5433.28923			

Table 8.9 Parameter estimates and their statistical significance

Variable	DF	Parameter estimate	Standard error	t value	Pr > |t|
Intercept	1	−14.14286	0.52585	26.90	<0.0001
Weeks	1	5.45824	0.07437	73.40	<0.0001

SAS Code for Regression Analysis

```
data fish;
input weeks fishweight; cards;
cards;
0 10.6091
27.225... ... ... ... ... ...
12 816492.96;
proc reg data=fish;
model fishweight=weeks;run;
```

8.5 Conclusion

The simple correlation and regression are arguably the most frequently used statistical tools by the fisheries biologist. In almost every data collected, correlation and regression are, at least, examined. But, in many cases, those tools are not correctly used possibly due to the lack of a strong conceptual base. This chapter describes simple correlation and regression, aiming at concept building—especially on the zero correlation with a strong nonlinear relationship. Most of the fisheries biologists are mainly concerned about regression estimates and their statistical significance, without diagnostics analysis. Hence, regression diagnostics is described elaborately with examples. The demo examples are executed in the SAS software. However, there are a lot of packages available to perform correlation and regression analysis. To begin with, Microsoft Excel is the immediate choice, but readily available regression diagnostics analysis is not available. The users can use free software, e.g. **R** software, for detailed analysis.

Exercise

1. The independent variable is also knows as
   (a) Estimated
   (b) Predictand
   (c) Regressor
   (d) Regressand
2. The dependent variable is also called as
   (a) Regressor
   (b) Independent
   (c) Regressand
   (d) All of the above
3. To determine the weight of a fish from its length is a
   (a) Correlation problem
   (b) Regression problem
   (c) Both of the above
   (d) None
4. If the value of any regression coefficient is zero, then the two variables are
   (a) Independent
   (b) Dependent.
   (c) Correlated.
   (d) None of the above
5. The regression line $y = a + bX$ will be decreasing if
   (a) $b > 0$
   (b) $b = 0$
   (c) $b < 0$
   (d) None of the above
6. The measure of the strength of the linear relationship is known as
   (a) Intercept
   (b) Correlation coefficient
7. When the observed and estimated values for the regression line $Y = a + bx$ are exactly the same, then the error sum of square is
   (a) Unknown
   (b) Zero
   (c) 1
   (d) -1
8. Correlation coefficient is independent of
   (a) Scale
   (b) Origin
   (c) Both of the above
   (d) None of the above
9. The correlation coefficient is the .... of two regression coefficient
   (a) Median.
   (b) Product.

(c) Geometric mean.
(d) Arithmetic mean.

10. When the regression line passes through the origin, then
    (a) The intercept is zero
    (b) The correlation is zero
    (c) The intercept is 1
    (d) The intercept is −1
11. The range of the correlation coefficient is
    (a) 0 to 1
    (b) −1 to +1
    (c) 0 to ∞
    (d) ∞ to +∞
12. The range of the regression coefficient is
    (a) 1 to 0
    (b) ∞ to 0
    (c) ∞ to +∞
    (d) −1 to ∞
13. The simple correlation coefficient is zero which means.
    (a) Weakly negative relationship.
    (b) Weakly positive relationship.
    (c) No relationship.
    (d) No linear relationship.
14. If $Y = -5X$ the regression of Y on X and $X = -0.2Y$, the regression of X on Y for the same data, then correlation coefficient between X and Y is
    (a) 0.5
    (b) −1
    (c) −0.2
    (d) None of the above
15. In correlation problem both the variables are
    (a) Fixed
    (b) Equal
    (c) Random
    (d) All of above
16. The length of fishes is measured in centimetre, and the weight is measured in gm; then the correlation coefficient is expressed as
    (a) cm/gm
    (b) gm/cm
    (c) Unit free
    (d) None of the above
17. The sign of correlation and regression coefficient is always the
    (a) Same
    (b) Different
    (c) Positive
    (d) Negative

18. In a simple linear regression model, if the independent variable changes by one unit, then the dependent variable changes by.
    (a) One unit
    (b) Zero unit
    (c) Intercept
    (d) Regression coefficient
19. What is the range of the values of r^2?
    (a) 0 to 1
    (b) −1 to 1
    (c) −1 to 0
    (d) All of the above
20. If the correlation coefficient between two variables is '1', then which of the following statement is true?
    (a) The relationship between them is perfectly linear and decreasing
    (b) The relationship between them is perfectly linear and increasing
    (c) There is no relationship between them
    (d) None of the above

Further Reading

Bhujel RC (2008) Statistics for aquaculture. Wiley-Blackwell, p 240 ISBN:978-0-813-81587-9
Draper NR, Smith H (2014) Applied regression analysis, vol 326. Wiley http://www.iasri.res.in/ebook/FET/Chap%202_Correlation%20and%20Regression.pdf
Gupta SC, Kapoor VK (2006) Fundamentals of mathematical statistics. Sultan Chand & Sons, Educational Publishers, New Delhi
https://cnx.org/contents/wAVi9myf@15/Linear-Regression-and-Correlat
http://www.fao.org/docrep/006/y4816e/y4816e0i.htm#TopOfPage
https://support.sas.com/documentation/cdl/en/statug/63962/HTML/default/viewer.htm#statug_reg_sect003.htm
http://www.iasri.res.in/ebook/FET/Chap%202_Correlation%20and%20Regression.pdf
Statistics for managers using Microsoft Excel, 2nd Edn (2004) Prentice-Hall.

Analysis and Interpretation of Weight-Length Data of Fish

9

Key Topics

- Importance of weight-length relationship
- Concept of weight–length relationship of fish
- Concept of condition factor and its uses
- Use of R software to determine weight–length relationship
- Interpretation of weight–length relationship

Most of the fisheries scientists regularly establish weight-length relationship. Typically, such work does not have merit for publication of any good scientific journal. However, fish species biodiversity is huge, and weight-length relationships are known for a restricted number of species. Establishing weight-length relationship is still required where such relationships are not previously known. To generate baseline information, establishing weight-length relationships is essential for the newly recorded fish species.

The basic idea is that the weight of a fish is a function of length with known form. Primarily, the weight-length relationship is used for estimating weight to a given length. Condition factor has almost been synonymous while investigating weight-length relationship of fish. It is used for comparing the condition of the fish of different sizes. The concept of weight-length relationship and 'condition factor' has been used in fisheries research since the twentieth century. There has been lot of confusion about correct application and interpretation of the relationship. Still the relationship has been widely applied in fisheries research.

It is useful to model aquatic ecosystems where observations are typically obtained as the number of specimen by length class, which has to be converted into estimates of biomass.

- It provides information on the changes in the well-being of the fish, which happens during their life cycle.

B. K. Das et al., *Concept Building in Fisheries Data Analysis*,
https://doi.org/10.1007/978-981-19-4411-6_9

- It is useful to diagnose the change during life cycle events like metamorphosis, growth and onset of maturity and spawning.
- The relationship can be used to differentiate taxonomic unit.
- This relationship can be used to set up yield equation and estimate landed number fish.
- Variation of general weight-length relationship provides indication to overall condition, which is usually analysed by using condition factor or ponderal index or coefficient of conditions derived from the parameters of weight-length relationship.
- Relative condition factor, which is derived from the parameter estimates of weight- length relationship, provides valuable information on maturation and spawning of fish, clues regarding the breeding season, fatness, etc.

Considering the aforementioned widespread applications, it is essential to learn how to establish weight-length relationship. This chapter provides guidelines for current fisheries researchers dealing with length-weight data so that they are able to give critical insights to length-weight data for better interpretation.

9.1 Concept of Weight-Length Relationship

The history of weight-length relationship is intertwined with the condition factor. Fulton (1904) had conceptualized the condition factor by using the concept of famous 'cube-law', i.e. the weight of fishes is said to increase approximately as the cube of the length. Accordingly the Fulton's condition factor is written as

$$K = \frac{100 \times W}{L^3} \tag{9.1}$$

where W is the whole body weight of fish in gram; L is the length of fish in cm; and the factor 100 is used to bring K close to unity. The above equation can easily be converted into the weight-length equation as

$$W = K \times L^3 \tag{9.2}$$

where the first mathematical form of weight-length relationship is based on the concept of 'cube-law'. Over the years, researchers have noticed that weight-length relationship deviates from 'cube-law' due to many factors, such as spatial, seasonal variation, food, nutrition, reproductive variations in condition, growth stanza, etc. Thus the concept of 'cube-law' has been changed, and the modern weight-length relationship is based on the concept of allometric growth and exponential nature of the variation of weight-at-length data. Using the above concept, the modern weight-length relationship (WLR) is represented mathematically as

$$W = aL^b \quad (9.3)$$

where W and L are variables as defined above and a and b are parameters.

9.2 Interpretation of Parameters

The parameter b in the WLR is the growth parameter. The functional regression value $b = 3$ implies the original 'cube-law'; it also describes the isometric growth, which would characterize a fish of unchanging body form and unchanging specific gravity. In other words, the rate of change in length and weight is the same. On the other hand, if b value is characteristically grater or less than 3, the condition is described as allometric growth. When $b < 3$, the growth is called negatively allometric; the rate of increase in length is faster than that in weight of fish; the shape of the fish tends toward flatness or attenuation when length increases. When $b > 3$, the growth is called positively allometric; the rate of increase in length is slower than that in weight; the fish becomes more spherical shape as length increases.

The parameter a in the WLR is called the weight-length factor. It resembles to the Fulton's condition factor. When $b = 3$, it is exactly the same as the Fulton's condition factor. So, a value in the equation describes the condition of the fish. Its fluctuations in value at any length unit are directly comparable with the fluctuations at any other length unit; this indicates relative changes in weight. However, an estimated form sample data provides very little information. In fact, condition of individual fish cannot be studied from a single sample. Investigation on individual variation from the general WLR has more importance than general WLR. It is frequently studied under the general name of 'condition', which is described below.

9.3 Condition Factor

Changes in individual fish condition are usually analysed by using a 'condition factor', which is also sometimes coined as 'coefficient of condition' or 'ponderal index'. The basic idea is to study the variation between observed weight and that expected from the observed length. Various types of condition factor have been used, but we will describe only two popular condition factors. The first one is the original Fulton's condition factor, and the other one is 'relative condition factor' derived from WLR. The computation of Fulton's condition factor is straightforward, by using the following formula:

$$K = \frac{W}{C \times L^3} \quad (9.4)$$

The value of c depends partly on the units used for weighing and measuring the fish. On the other hand, the 'relative condition factor' is computed from the formula as given below:

$$K_R = \frac{W}{\widehat{a}L^{\widehat{b}}} \tag{9.5}$$

It is clear from the above formula that the relative condition factor is the ratio of observed weight to the predicted weight. The following two-step strategies are followed to compute the relative condition factor.

- Determine the WLR from the data and predict the weight for each observed length.
- Compute the relative condition factor by dividing the observed weight by the predicted weight obtained as in step 1.

Differences in condition factor can be used to measure various biological features such as flatness, suitability of environment, and gonad development. The fish does not obey the cube-law in its WLR. Thus length itself and factors correlated with affect the values of *K*. So a plot of condition factor against length may be useful to detect changes in the condition of fish during its growth. Further, age, sex or maturity may affect the values of condition factor. Differences of mean of K for fish from different environment can be compared. For some species seasonal changes of *K* may provide clues to determine spawning season.

In view of the list of widely different factors that can affect condition factor, it is not surprising that interpretation of condition factor is difficult and often leads to erroneous results.

9.4 Determining WLR from Observed Data

When length-weight data on fairly large number of specimens have been collected, it is easy to determine the WLR. Estimation of two parameters, e.g. *a* and *b*, is required to establish the WLR. After log transformation the mathematical formula of WLR becomes

$$\text{Log}(W) = \log(a) + b \times \log(L) \tag{9.6}$$

If the both the length and weight data are transformed into their logarithm, the estimation process is similar to fitting simple linear regression. In practice, a simple linear regression is fitted after taking log transformation of the original data. It is noted that *b* value will be the same as obtained from the transformed variables.

However, one has to take antilog transformation of the intercept to get the actual a value. Suppose after transformation the intercept is obtained as a_0. Then a value is obtained as

$$A = e^{a_0}$$

All these can easily be carried out in any software, including SAS, SPSS, R, Excel, etc. In the next section, we will demonstrate to determine WLR using a real data example, by R software.

9.4.1 Analysis of Weight-Length Data Using R Software

It is prerequisite that the readers have some basic knowledge and operational skill on R software. This section demonstrates the analysis of real length-weight data of a fish species.

Brief Description of the Data
The length-weight data of large number (1100) of specimens of certain species were collected over different months. The data were grouped by sex. The snapshot of data in Excel is shown in Table 9.1. We will analyse this data according to the aforementioned theory.

The objectives here are as follows:

- To establish the WLR
- To determine the 'Fulton's condition factor' and 'relative condition factor'
- To change detection of condition by using graphs

Note *At first, make suitable data entry in your favourite spreadsheet applications, e.g. Microsoft Excel, and then convert the data sheet to a .csv file. Thereafter, follow the R code given in the following box. Modify the code as per the data set.*

Table 9.1 An abridged dataset of weight-length of a fish species

Month	Place	Length (mm)	Weight (gm)	Sex
March	Farakka	184	62	M
March	Farakka	260	196	M
March	Farakka	230	123	F
March	Farakka	170	64	F
March	Farakka	...	...	...

Step 1: Convert the Excel data sheet to a .csv file to bring the data into R.

```
>dat=read.csv(file.choose(), header=TRUE) ### reading .csv file
   and assgn it into 'dat'
#### WLR ###
>fit=lm(log(Weight)~log(Length),data=dat) ### estimating WRL
   after log transformation
>summary(fit) ### to display summarized results, here R-squared
   value can be seen
>b= fit$coeff[2] #### Extracting b value
>a=exp(fit$coeff[1]) #### Extracting a value
######## Plotting WRL #####
>plot(dat$Length,dat$Weight,xlab="Length#$#$(mm)",
ylab="Weight(gram)",
   main="") #### plotting the points
curve(a*x^b,0,max(dat$Length), col="red",add=TRUE) ### adding
   the WLR
########## Placing the equation and R-square value in the plot
>text(100,1000,expression( W==5.04%*% 10^-6 %*%L^3.11))
>text( 100,800, expression (R^2==0.99))
```

Note
Interpretation: *The* R^2 *value of* 0.99 *is very high; hence, the WLR fitted well for the data. We found the b value is* 3.11 *which is greater than* 3. *So the growth of the fish species is not isometric. Positive allometric growth is observed for the species. This means that the growth rate of weight is higher than that of length. As the fish grows, it becomes more of spherical shape. This is probably due to development of gonad and other factors.*

Note ***Graphical presentation:*** *In most of the software, the weight-length relationship is established by using log transformation of weight and length. Thereby, log-log plot is readily available from the software. But, it is recommended to present the relationship in terms of the observed weight and length as shown in* Fig. 9.1.

9.5 Determine Fulton's and Relative Condition Factor

The computation of Fulton's and relative condition factor is very easy to compute, using Eqs. (9.4) and (9.5). The same can be computed by using Excel. However, the following R-code is provided.

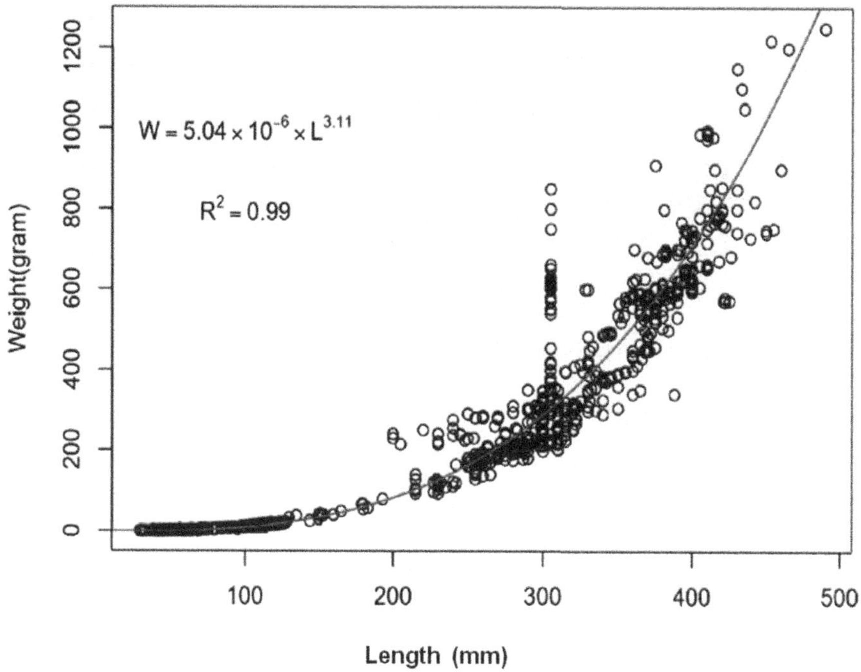

Fig. 9.1 The weight-length relationship of the studied fish species

```
######## Computing Condition factor
K.f=(dat$Weight/dat$Lenght^3)*1000 ####### Fulton's Condition
   Factor
K.R= dat$Weight/(a*dat$Length^b) #### Relative Condition factor
########Plotting condition factors over months
>m.levels=c("Jan", "Feb", "Mar", "Apr", "May", "Jun", "Jul", "
   Aug", "Sep", "Oct", "Nov", "Dec")
>f.levels=c("January", "February", "March", "April", "May", "
   June", "July", "August", "September", "October", "November",
   "December")
>F.month=factor(dat$Month, levels=f.levels, labels=m.levels)
>boxplot(K.f~F.month,xlab="Month", ylab="Fulton's#$Condition
   #$Factor")
>boxplot(K.R~F.month,xlab="Month", ylab="Relative#$Conditon
   #$factor")
```

Note

Interpretation: *The condition factors drop at the month of July and regain quickly after that (see* Fig. 9.2*). It indicates that July may be the onset of spawning of the*

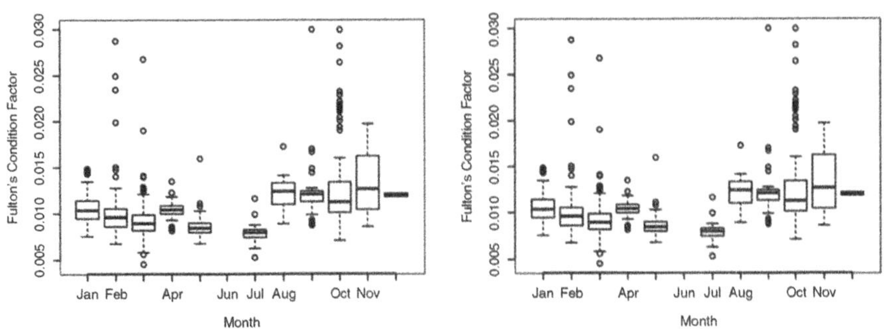

Fig. 9.2 Fulton's and relative condition factor for different months

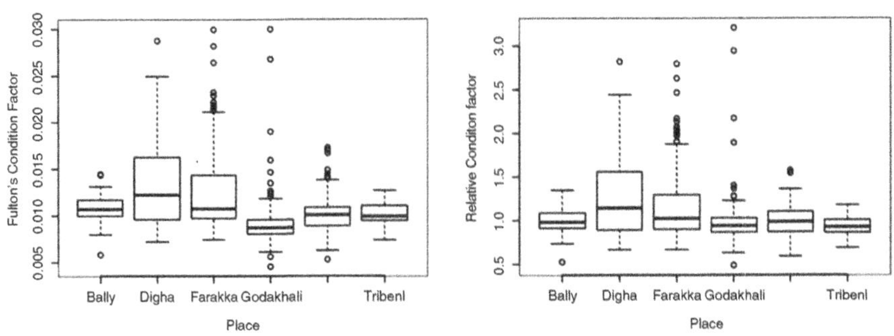

Fig. 9.3 Fulton's and relative condition factor for different places

species. If it relates gonadal condition, the subsequent months are the spawning season.

Plotting Condition Factors for Different Places

```
>boxplot(K.f~dat$Place,xlab="Place", ylab="Fulton's#$Condition
   #$Factor")
>boxplot(K.R~dat$Place,xlab= "Place",ylab= "Relative
   #$Condition#$factor")
```

Note

Interpretation: *Relative condition factor against places (see* Fig. 9.3*) indicates that it is higher than unity at Digha. This means observed weight is more than the expected weight at Digha. The fatness of fishes at Digha is relatively higher than those in other places. There could be many reasons.*

9.5.1 Plotting Condition Factors Against Length

```
>scatter.smooth(dat$Length,K.f,col= "red",xlab= "Length(mm)",
ylab="Fulton's#$condition#$Factor")
>scatter.smooth(dat$Length,K.R,col= "red",xlab= "Length(mm)",
   ylab="Relative#$Condition#$Factor")
```

Note

Interpretation *As such, it does not reflect any abnormal changes in the condition factor against length (see* Fig. 9.4*). In some case sudden drops at particular length may be an indicative to spawning.*

9.6 Conclusion

1. When gathering length-weight data, strive to select the size range for which the relationship does not alter. It is recommended not to include to early juveniles those which have not yet attained adult body shape. If required separate relationship for juveniles may be established. Also do not include those specimens that have distorted their shape.
2. Make sure that there is no bias introduced by selective gear. For example, gill net of only one mesh size will capture only the fishes around the targeted size. Specimens from entire range of length will not be obtained.
3. Make sure to include about equal numbers of randomly selected small, medium-sized and large specimens. There is no need to measure large numbers of abundant size specimen.
4. Try to collect length-weight data for as many months as possible, to study seasonal variation.
5. Test for differences between sexes. If such differences are significant, present separate relationships for males, females and both sexes.

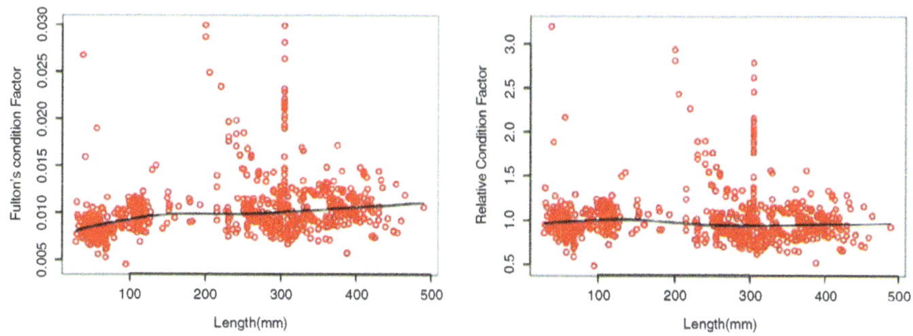

Fig. 9.4 Fulton's and relative condition factor against length

6. When comparing your results with published WLR, make sure data were collected during the same season in the same adjacent areas.
7. Detect outliers before finalizing WLR. Use residual diagnostic plots to detect outliers, and re-estimate the WLR.
8. While presenting WLR results, make sure to include the following items.
   (a) The number of specimens measured
   (b) Range
   (c) Type of length measurements (girth, total length, standard length, etc.)
   (d) Units of measurements
   (e) Parameters a and b with their 95% confidence interval
   (f) Coefficients of variation
   (g) WLR graph or scatter plot superimposed with estimated WLR

Exercise

1. Which of the following equation is used for weight-length relationship of fish?
   (a) $W = aL^b$
   (b) $\text{Log(W)} = \log(a) + b\ X \log(L)$
   (c) Both of the above
   (d) $W = a + bL$
2. Which of the following weight-length relationship indicates isometric growth of fish?
   (a) $W = a + 3\ L$.
   (b) $W = aL^3$.
   (c) $W = 0.1\ X\ L$.
   (d) None of the above
3. In WLR equation, $W = aL^b$ is known as the.
   (a) Condition factor
   (b) Coefficient of condition
   (c) Ponderal index
   (d) All
4. In WLR if $b = 3$, it indicates that
   (a) Rate of change of length is more than that of weight
   (b) Rate of change of length is less than that of weight
   (c) Rate of change of length and weight is the same
   (d) None
5. For developing weight-length relationship, what is the standard unit of measurement of length and weight?
   (a) Length in mm and weight in milligram
   (b) Length in cm and weight in gram
   (c) Length in mm and weight in kg
   (d) Length in m and weight in kg
6. Suppose a weight-length relationship $W = aL^b$ is developed to predict the weight of a fish species by using its length as the explanatory variable. Weight

was measured in gram for a sample of 100 fish from a fish market. Now, suppose that the weight of fish is converted to kilogram. The impact of this conversion is/are:
   (a) The magnitude of a will change
   (b) The magnitude of *b* will change
   (c) Both *a* and *b* will change
   (d) Neither *a* nor *b* will change
7. In a weight-length relationship of a fish species, *b* value is less than 3. Then which of the following is correct?
   (a) The species has had negative allometric growth
   (b) The species grows isometrically
   (c) The species has had positive allometric growth
   (d) None of the above
8. Suppose that in a WLR for a fish species, the *b* values is estimated as **3.5**. Which of the following statements is correct?
   (a) The species growth allometry is negative
   (b) The species growth allometry is positive
   (c) The rate of increase in length is slower than that in weight
   (d) Both of (b) and (c)
9. Specify whether the statement 'Selective gear is ideal to determine weight length relationship' is true or false.
   (a) True
   (b) False
10. Condition factor provides information on the
   (a) Age of fish
   (b) Growth rate of the fish
   (c) Size of the fish
   (d) Well-being of the fish

Further Reading

Froese R (2006) Cube law, condition factor and weight–length relationships: history, meta-analysis and recommendations. J Appl Ichthyol 22(4):241–253. https://doi.org/10.1111/j.1439-0426.2006.00805.x

Le Cren ED (1951) The length–weight relationship and seasonal cycle in gonad weight and condition in the perch (Perca fluviatilis). J Anim Ecol 20:201–219

Safran P (1992) Theoretical analysis of the weight–length relationship in fish juveniles. Mar Biol 112:545–551

Some Multivariate Analysis Techniques for Fisheries Data

10

Key Topics
- Overview of different multivariate analysis techniques
- Principal component analysis
- Multiple regression analysis

Biological, physical, behavioural, social and educational phenomena are often characterized by multiple factors or variables. Therefore, any systematic attempt to understand them requires the examination of multiple dimensions that are usually interrelated. The subject of multivariate analysis deals with the statistical analysis of the data collected on more than one variable. These variables may be correlated with each other. In practical situation, most data collection schemes or designed experiments result in multivariate data. A few examples of such situations are given below.

- During a survey of households, several measurements on each household are taken. These measurements, being taken on the same household, will be dependent. For example, the education level of the head of the household and the annual income of the family are related.
- In exploratory survey of aquatic ecosystem, several measurements (number of habitat variables, water chemistry data, etc.) are recorded to characterize the water body. It can be contemplated that those variables are interrelated due to ecosystem function.
- In a designed experiment conducted in a research and development centre, various factors are set up at desired levels, and a number of response variables are measured for each of these treatment combinations. Due to dependence among responses, it may be more meaningful to analyse response variables simultaneously.

B. K. Das et al., *Concept Building in Fisheries Data Analysis*,
https://doi.org/10.1007/978-981-19-4411-6_10

Generally, data are analysed by taking one variable at a time. But in all above examples, collected data are multivariate in nature. The inferences drawn by analysing the data for each of the variables may be misleading. Therefore, the data on several variables should be analysed using multivariate analytical techniques. Various statistical methods for describing and analysing these multivariate data are available, such as multivariate analysis of variance (MANOVA), discriminant analysis, factor analysis, principal component analysis (PCA), cluster analysis, canonical correlation analysis (CCA), multidimensional scaling, correspondence analysis, canonical correspondence analysis, multiple regression analysis, etc. However, this chapter describes important and frequently used classical multivariate techniques, viz. PCA and multiple regression analysis and cluster analysis. These have also been demonstrated using R and SAS Software.

10.1 Principal Component Analysis

The purpose of PCA is dimension reduction. Essentially, it derives a small number of linear combinations (principal components) of observed variables that retain as much information in the original variables as possible. Often a small number of principal components can be used in place of the original variables to explore relationship among variables, regression, grouping or clustering of individuals or observation, etc. PCA can also be viewed as a technique to remove multicollinearity in the data. In this technique, we transform the original set of correlated variables to a new set of uncorrelated random variables. These new variables are linear combinations of the original variables and are derived in decreasing order of importance so that the first principal component accounts for as much as possible of the variation in the original data.

It is quite likely that first few principal components account for most of the variability in the original data. If so, these few principal components can then replace the initial variables in subsequent analysis, thus reducing the effective dimensionality of the problem. An analysis of principal components often reveals relationships that were not previously suspected and thereby allows interpretation that would not ordinarily result. However, PCA is more of a mean to an end rather than end in itself because this frequently serves as intermediate steps in much larger investigations by reducing the dimensionality of the problem and providing easier interpretation. It is a mathematical technique, which does not require the user to specify the statistical model or assumption about distribution of original variables. It may also be mentioned that principal components are latent variables, and often it is not possible to assign physical meaning to them. Further, since PCA transforms original set of variables to new set of uncorrelated variables, it is worth stressing that if the original variables are uncorrelated, then there is no point in carrying out the PCA. It is important to note here that the principal components depend on the scale of measurement. Conventional way of getting rid of this problem is to use the standardized variables with unit variances.

10.1.1 PCA Workflow

1. Arrange data in the form of a matrix or table.
2. Compute mean and standard deviation of each of the variables.
3. Standardize the variables by subtracting mean from observed value, and thereafter divide by standard deviation.
4. Examine the pairwise correlation (if correlated, go for PCA; otherwise stop).
5. Perform PCA using software and find principal components.
6. Choice of number of components; careful attention to be taken to choose the number of component. The basic criterion is the variability explained by the number of components.
7. Producing useful maps of variable and sample.
8. Understanding how these maps are constructed will maximize the information we can gather from them.

10.1.2 Real Data Example

Table 10.1 is reproduced from a secondary survey data on 14 rivers. There were seven variables to characterize the river condition which are assumed to have the effect on fish species richness. Values of the seven habitat variables were recorded. The units of each variable are different and omitted due to demo purpose.

10.1.2.1 Questions

- Are the variables interrelated among themselves?
- Can we explain total variability in terms of few variable?
- How the variables are structured in reduced dimension?
- Whether grouping of rivers is possible according to habitat and climatic variability?

10.1.2.2 Step-by-Step Analysis

PCA is the prudent choice to get solution of aforementioned questions. The workflow details is described as follows:

Examining interdependence: The immediate choice is to examine the pairwise correlation, which can be computed in any software. The results of pairwise correlation are shown in Table 10.2. Note that few variables are intercorrelated. For example, high positive correlation of 0.87 is observed between water temperature and latitude. It is worthy to carry out PCA for dimension reduction.

Choosing the number of components: Execution of PCA in any software will provide output of eigenvalues in order of magnitude. The following R script is provided to generate scree plot under R software.

Table 10.1 Fish species richness and climato-hydrological variables of 14 major rivers of India

Temp. range	Rainfall	Discharge	Sediment	Basin area	Latitude	Water temp.	Richness
9.25	2280.34	16,180.80	567,544.60	187,068.20	18,793.17	23.13	221
3.00	734.51	93.11	207.97	87,902.25	16,330.52	26.9	95
13.74	922.57	15,631.48	1458.81	860,993.80	77,268.06	26.2	265
7.00	719.45	118.85	51,404.37	312,607.90	28,248.80	28.2	91
4.75	736.21	250.61	1035.14	258,821.3	22,646.44	29.5	113
9.51	1124.61	1169.5	12,078.14	141,579.4	18,238.96	27.1	253
6.5	680.77	94.84	4677.35	34,833.73	11,912.42	26.4	31
7	688.65	843.33	6123.5	98,855.3	29,308.93	25.3	57
4.5	1621.81	226.46		5395.11	6516.28	27.8	115
6.75	734.51	14.59	176.6	21,677.04	6729.77	27.9	34
9.75	584.79	266.69	23,334.58	65,162.84	15,205.48	28.2	56
6.32	827.94	549.54	207.97	25,882.13	14,321.88	15.3	54
5	453.94	212.81	12,078.14	24,099.05	4602.57	21.5	57
11.75	1091.44	14.55	6123.5	23,388.37	13,243.42	28.5	56

Table 10.2 Pairwise correlation among the climato-hydrological variables. (1) Temperature range; (2) Rainfall; (3) Discharge; (4) Sediment; (5) Basin area; (6) Latitude; (7) Water temperature

	1	2	3	4	5	6	7
1	1.00	0.07	0.03	−0.29	0.10	0.19	0.14
2	0.07	1.00	0.25	0.36	0.48	0.36	0.46
3	0.03	0.25	1.00	0.42	0.21	−0.01	0.11
4	−0.29	0.36	0.42	1.00	0.39	0.58	0.58
5	0.10	0.48	0.21	0.39	1.00	0.39	0.20
6	0.19	0.36	−0.01	0.58	0.39	1.00	0.87
7	0.14	0.46	0.11	0.58	0.20	0.87	1.00

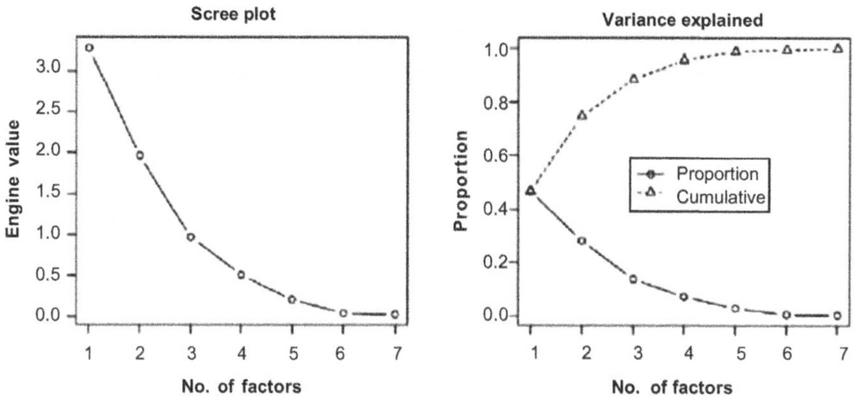

Fig. 10.1 Variance explained against the number of principal components

```
>pc.fit=princomp (x, cor=TRUE)
>plot(pc.fit)
```

Note: *Here x is data matrix;* ***cor=TRUE*** *is used to standardize the data, and analysis is performed on the basis of correlation matrix. Otherwise, bizarre results may be produced due to different measurement units.*

The thumb rule is to select the number of components having eigenvalues more than one. However, choice depends on the data interpretation as well. Sometimes eigenvalue of low magnitude can also be selected, depending upon the objective of study. Figure 10.1 shows that three principal components/factors (eigenvalue ≥ 1, a thumb rule for number of factor selection) could be selected to explain reasonable amount of variability (88.7%) for further analysis. However, the variables were not well segregated in defining the extracted factor in a meaningful way. Though fourth factor (eigenvalue = 0.51) explain very little amount of variation (7.3%), factors interpretation became meaningful with the inclusion of this factor. In addition, such inclusion ascertained 96% of total variability present in the included climato-hydrological variables in the study, which restricts us to retain only four factors.

Structural association of variable with principal components: PCA analysis in any software also provides factor loading values. Sometimes, structural association is not discernible with the actual loading. Then loading of rotation (varimax, etc.) can be used for better separation of variable with the principal components. The first concern in interpreting the principal components and loading to derive latent factors that have practical utility. If it is not possible, it is better explain the data in terms of principal components itself. In our example we have derived synthetic variable in terms of river habitat in the context of fish species richness as given below.

With a view to extract interpretable factors from complex climato-hydrological interaction in river system, we took the help of factor loading patterns of each variable under study. Table 10.3 represents the loading patterns of each variable for unrotated and varimax rotated state. It is observed that variables are not well separated in terms of loading. For example, sediment load and rainfall have reasonable contribution ($\geq$0.5) to Factor 1 and Factor 2 (see Table 10.3). Moreover, these factors do not represent any particular climato-hydrological feature. It does not provide any factor interpretation in a realistic manner. When these factors are rotated orthogonally by varimax transformation, we clearly observe the variable segregation. Factor 4 is dominated by the influence of temperature stability (loading = 0.906), whereas Factor 3 is explaining the variability of the water temperature (loading = 0.992).

This is clearly depicted in Fig. 10.2, in which gradient of temperature stability towards Factor 4 and of water temperature towards Factor 3 is almost zero.

Moreover, extent of contribution (length of the line in figure) is very near to unit circle. This leads us to define Factor 4 as temperature stability and Factor 3 as water temperature. In a similar way, we identify (see Fig. 10.2) that basins are main contributory variables for Factor 1. As these two variables simultaneously represent regional surface area of the river, we call the Factor 1 as 'regional surface area'. On the basis of gradient (see Fig. 10.2) and loading (see Table 10.3), we observed that Factor 2 is the integrated effect of rainfall, discharge and sediment load which are very much related to river hydrology. Rainfall is one of the renewal sources of water and influence greatly on water retention of the river. Discharge and sediment load also play important role in water retention, as depth and area are confounded in discharge and sediment load. Though there are other parameters (e.g. water abstraction) for exact formulation of water retention, we call this confounded effect of these three variables representing Factor 2 as 'water-retentive potential' of the river.

Distribution pattern of samples (here rivers) Few computations are required to find distributional pattern of each sample (here 14 rivers). From the software, factor scores have to be generated for each sample. Then plot these scores in two dimensions.

The sample R code is as follows:

Table 10.3 Loadings of each variables to the extracted factor

Variables	Unrotated factor				Varimax rotated factor			
	1	2	3	4	1	2	3	4
Temperature range	0.733	−0.219	0.106	−0.634	0.391	0.149	0.062	0.906
Rainfall	0.500	0.748	0.277	0.025	−0.029	0.936	0.047	0.094
Discharge	0.948	0.227	−0.095	0.099	0.370	0.864	−0.155	0.247
Sediment	0.524	0.787	0.090	0.079	0.028	0.943	−0.135	0.032
Basin area	0.821	−0.493	−0.075	0.244	0.971	0.036	0.092	0.173
Latitude	0.800	−0.544	−0.143	0.145	0.954	−0.043	0.0287	0.254
Water temp.	0.003	−0.380	0.915	0.111	0.054	−0.080	0.992	0.047

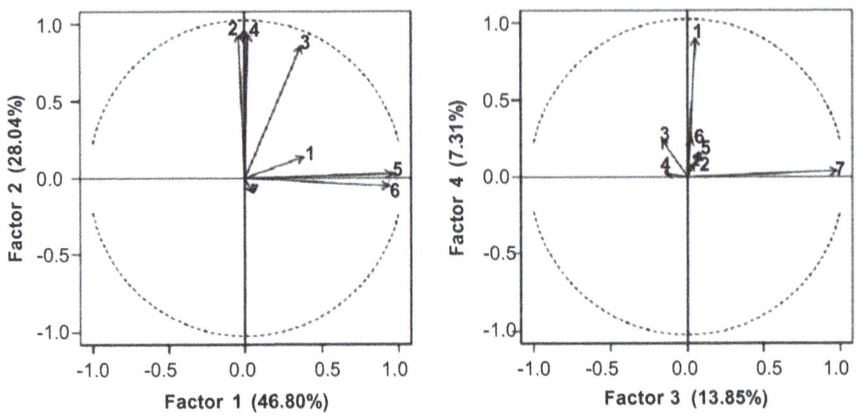

Fig. 10.2 Structural dependence of variables with principal component

R-Code

```
>pc.fit=princomp(x,cor=TRUE,scores=TRUE )
>plot(pc.fit$scores[,1],pc.fit$scores[,2],xlab="PC1",ylab="PC2",
type="n")
>abline(h=0,v=0,lty =2)
>text(pc.fit$scores[,1],pc.fit$scores [,2],c(1:14))
```

Figure 10.3 shows the distribution pattern according to two principal components. Note that river 1 is totally different from others. The component 1 is separating 9, 13, 12 and 1 from others, while the rest of the rivers is forming another group according to climato-hydrological parameters of the rivers.

10.1.2.3 PCA Through SAS

The following steps of SAS may be used for performing PCA: the PROC PRINCOMP can be used for performing PCA. Raw data, a correlation matrix, a covariance matrix or a sum of squares and cross products (SSCP) matrix can be used as input data. The data sets containing eigenvalues, eigenvectors and standardized or non-standardized principal component scores can be created as output. The basic syntax of PROC PRINCOMP is as follows:

```
title 'PCA_by_using_SAS';
data pca;
input Temp\_Ran Rainfall Discharge Sediment Bas\_Area Latitude
Water
```

(continued)

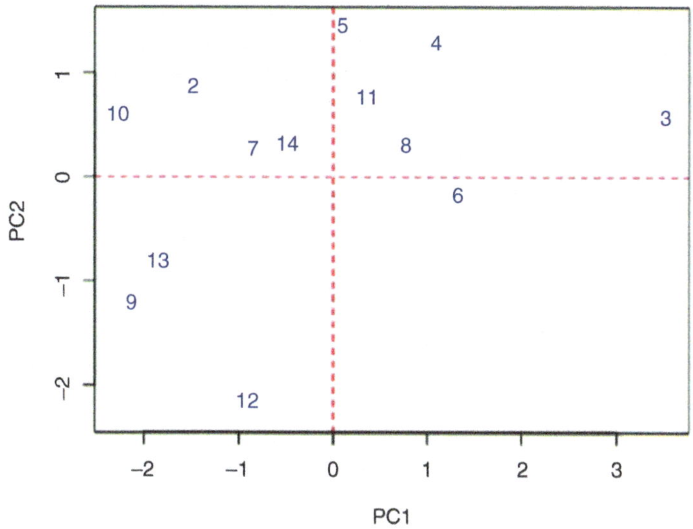

Fig. 10.3 Score distribution pattern of 14 rivers against two main components

```
\_Temp;
cards;
\verb|odsrtffile='E:\SAS#$data\pca.rtf';|
proc princomp corr out=new;
var Temp\_Ran Rainfall Discharge Sediment Bas\_Area Latitude
Water\_Temp;
proc print data=new ;
var Temp\_Ran Rainfall Discharge Sediment Bas\_Area Latitude
Water\_Temp; prin1
prin2;
proc corr data=new ;
var Temp\_Ran Rainfall Discharge Sediment Bas\_Area Latitude
Water\_
Temp; prin1 prin2;
run;
```

The PROC PRINCOMP and RUN are a must. However, the VAR statement listing the numeric variables to be analysed is usually used along with PROC PRINCOMP statement. If the DATA=data set is TYPE=SSCP, the default set of variables does not include intercept. Therefore, INTERCEPT may also be included in the VAR statement. The following options are available in PROC PRINCOMP.

Data Set Specification

1. DATA=SAS-data-set: names the SAS data set to be analysed. This data set can be ordinary data set or a TYPE=CORR, COV, FACTOR, UCORR or UCOV data set.
2. OUT=SAS-data-set: creates an output data set containing original data along with principal component scores.
3. OUTSTAT-SAS-data-set: creates an output data set containing means, standard deviations, number of observations, correlations or covariances, eigenvalues and eigenvectors.

Analytical Detail Specification

1. CORR: computes the principal components from the correlation matrix. The default option is computation of principal components using a correlation matrix.
2. N=: the non-negative integer equal to the number of principal components to be computed.
3. NOINT: omits the intercept from the model.
4. PREFIX= name: specifies a prefix for naming the principal components. The default option is PRIN1, PRIN2, ...
5. STANDARD (STD): standardizes the principal component scores to unit variance from the variance equal to corresponding eigenvalue.
6. VARDEF=DF|N|WDF|WEIGHT: specifies the divisor (error degree of freedom|number of observations|sum of weights|sum of weights-1) in calculating variances and standard deviations. The default option is DF.

Besides these options NOPRINT option suppresses the output. The other statements in PROC PRINCOMP are as follows:

By variables: obtains the separate analysis on observations in groups defined by variables.

FREQ statement: It names a variable that provides frequencies of each observation in the data set. Specifically, if n is the value of the FREQ variable for a given observation, then that observation is used 'n' times.

PARTIAL statement: used to analyse for a partial correlation or covariance matrix.

VAR statement: lists the numeric variables to be analysed.

WEIGHT statement: If we want to use relative weights for each observation in the input data set, place the weights in a variable in the data set, and specify the name in a weight statement. This is often done when the variance associated with each observation is different, and the values of the weight variable are proportional to reciprocals of the variances. The observation is used in the analysis only if the value of the WEIGHT statement variable is non-missing and greater than zero.

The other closely related procedures with PROC PRINCOMP are the following:

PROC PRINQUAL: It performs a PCA of a qualitative data.
PROC CORRESP: Performs correspondence analysis, which is a weighted PCA of contingency tables.

10.2 Multiple Regression Analysis

The multiple regression is used when data is available on three or more measurement variables; one of the measurement variables is the dependent (Y) variable. The rest of the variables are the independent (X) variables or explanatory variables, which are thought to have effect on the dependent variables. The purpose of a multiple regression is to find an equation that best predicts the Y variable as a linear function of the X variables. There are two basic uses of multiple regression: (1) for prediction of dependent variables based on independent variable, i.e. estimation of an unknown value of dependent variables based on known set of values of independent variables, and (2) understanding the causes. The latter use of multiple regression is to try to understand the functional relationship between dependent and independent variables, which might be useful to characterize the variation in the independent variable. However, one has to be very careful to use multiple regression to understand cause-effect relationship. Otherwise, the results of multiple regression may be misleading. In fact, results of multiple regression are more of a suggestion than for testing of hypothesis.

Hypothesis
When the multiple regression is used for hypothesis testing, the main null hypothesis is that there is no linear relationship between dependent variables (Y) and a set of independent variables (X).

10.2.1 Working Principle

The basic concept of multiple regression is built up on the basis of some basic assumptions as follows:

- The functional relationship between dependent and a set of independent variables is linear.
- The underlying distribution of error is normal for hypothesis testing.

Mathematically, the functional relationship between dependent and independent variables is written as follows:

$$Y = b_0 + b_1 X_1 + b_2X_2 + \ldots + bpXp + e \tag{10.1}$$

where b_0 is the intercept; $b_1, b_2, \ldots, b_p$ are the regression coefficients corresponding to each variables; and e is the random error—typically assumed to follow normal distribution.

The principle of multiple regression is to find the values of regression coefficients and intercept, which minimize the sum of square deviations of observed and expected values of Y. The multiple correlation—usually denoted as R^2 and commonly known as 'Coefficient of determination'—is used as the criterion for fitting the relationship to the data. When the purpose of multiple regression is prediction, the set of known values of independent variables is plugged into the fitted equation to obtain the predicted value of the dependent variables. When the purpose of multiple regression is understanding functional relationship, the effect of each variable is tested based on the estimated value of regression coefficient under the condition that other independent variables are kept constant. But the magnitude of estimated regression coefficients does not tell anything about the relative importance of corresponding independent variables. To overcome this, standardized regression coefficient is used that will tell something about the relative importance of different variables. For example, the independent variables with bigger magnitude (excluding sign) standardized regression coefficient have stronger relationship with dependent variables (Y) as compared to others. It is very easy to compute in post hoc analysis in any standard statistical packages. Typically, every software provides output of the estimates of regression coefficients and its corresponding estimated standard error. The standardized regression coefficient is then computed by dividing the estimate by the corresponding standard errors.

Note

Some Warnings

- *It is easy to do software-based multiple regression analysis with multidimensional data to get impressive looking output. However, users are very often sceptical of usefulness of multiple regression—especially in the case of variable selection. For the same data, different researchers choose different set of independent variables to come out with different results for the same data. It is very important to select those independent variables that cause the variability in the dependent variables. There are objective approaches like stepwise multiple regression, but it may so happen that the included variables do not have plausible explanation for Y.*
- *Before conducting multiple regression analysis, make sure that data is not too short, and multiple regression will fail if the number of data points is less than or equal to the number of independent variables.*
- *The multiple regression is based on some assumptions. The results will be misleading if those assumptions do not hold true. One crucial assumption is that the error is normally distributed—which is difficult to hold true for real-life data in the inland fisheries—especially for hypothesis testing. In such situation, rather than rigorous hypothesis testing, understanding functional relationship should be given critical insight for exploring cause-effect relationship. Another assumption is linear relationship between dependent variables and independent*

variable. Exploratory analysis may be conducted to ensure the relationship pattern before conducting multiple regression analysis.

- *Very often, multicollinearity problem misleads the interpretation of results of multiple regression. It occurs when two or more independent variables are highly correlated with each other. When the goal is prediction, then multicollinearity is not that important. However, if the goal is to understand causes, multicollinearity may confuse in identifying the cause-effect relationship. Before doing multiple regression, one should check the correlation between each pair of independent variables. If two are highly correlated, it is recommended to pick just any one of them.*

10.2.2 Workflow of Multiple Regression Analysis

Step 1: Fisheries practitioners have a tendency to go directly to the software, with the expectation that software will do the wonders for them. But this is totally a wrong strategy. Cleaning and pre-processing is very much essential; otherwise it may lead to wrong inferences. For example, there may be outlying observation, missing data and error due to data recording, which need to be handle with care—some righteous manipulation (deleting of outlier, etc.) might have to do.

Step 2: In real-life data, it is very often required to create new variables. This step may appear to be easy but can take longer time than anyone can think of it. It is very rare that every variable collected is in right form. If required, create new indices, derived variable (like conversion frequencies in percentage) to get variables in their final form of data interpretation.

Step 3: Basic exploratory data analysis is required to examine the underlying assumption for the analysis. It is pre-requisite to check the distribution of the variables intended to use; histogram plot of each variable may help to identify the distributional pattern. As multiple regression assumes linear relationship, thereby pairwise bivariate plot might have to do the examination of the relationship. If the goal is to structure relationship between independent and dependent variables, multicollinearity diagnostics are essential part before execution of multiple regression analysis.

Step 4: Execute the initial multiple regression analysis, and this will not be the final results. It is needed to explore further with the fitted model and refine it. Finalize the result based on post-analysis diagnostics (e.g. observed vs. predicted plot, residual plot, outliers, influential points, etc.) of regression model.

10.2.3 Real Data Example

The workflow of multiple regression is illustrated by the real data presented in Table 10.1. The data was emanated from secondary sources to investigate the effect of climato-hydrological factors on the fish species richness in riverine system.

Hypothesis

- How climato-hydrological factors influence fish species richness?
- What are the most influential factors of riverine fish species richness?

10.2.3.1 Systematic Analysis

The following procedure is followed to examine the above hypothesis.

From the question it is immediate that fish species richness is dependent, and the rest of the variables are independent variables. Though this list of influential factors is not exhaustive, the independent variables have been selected from experts. In addition, different variables have different measurement units. So, it is better to standardize the variables using the following formula:

$$X_{\text{standardize}} = \frac{X - \overline{X}}{\text{sd}} \tag{10.2}$$

Run initial multiple regression analysis. The R-code to run the multiple regression is as follows:

```
>dat=read.csv("rspecies.csv", header = TRUE) ### Read the data
file
>fit<-lm(SpeciesRichness$\sim$TempRange+Rainfall+Discharge
+Sediment+
BasinArea+Latitude+WaterTemp,data=dat) \#### Fitting multiple
regression
>error<-resid(fit) \##### computing residual
>hist(error) ######## Histogram plot of the residuals
```

Note that the error has been computed from the fitted model, for checking normality assumption of the error distribution. The histogram plot (see Fig. 10.4) depicts the distribution of error, which immediately suggests that error distribution is skewed, at least, not normal.

This directs us to create new variables to make error distribution normal distribution. Since it is positively skewed, a log transformation (log10(SpeciesRichness+1) is tried, which resulted in following error distribution after fitting multiple regression model. Note that the error distribution is now somewhat close to the normal distribution (see Fig. 10.5). It is pertinent to mention here that there are other means to examine the normality, such as Q-Q plot, cumulative distribution plot, etc. The results of the multiple regression are produced in Table 10.4.

10.2.3.2 Interpretation of R^2

The results show that coefficient of determination (here multiple $R^2 = 0.73$ and adjusted $R^2 = 0.43$), which is not statistically significant (p-value $= 0.15$). The interpretation of R-squared is that the multiple regression explained 43% variability of fish species richness by the independent variables, and it is not statistically

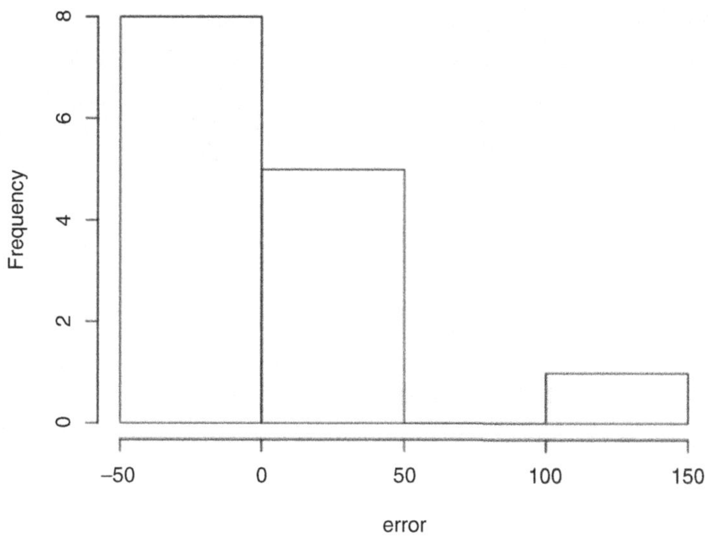

Fig. 10.4 Distribution of errors for non-transformed species richness

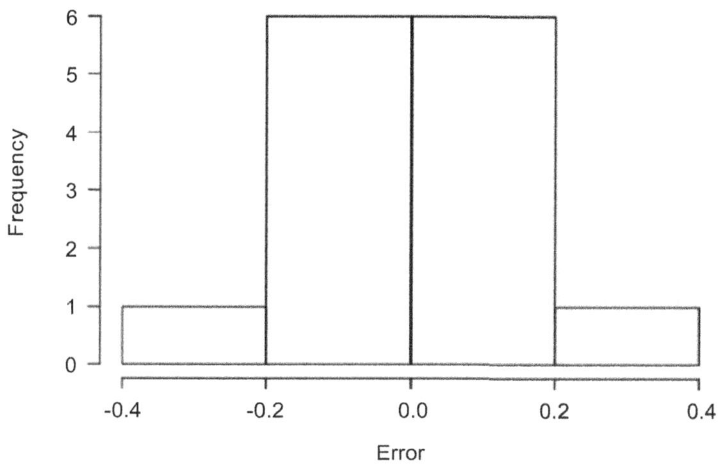

Fig. 10.5 Distribution of errors for log-transformed species richness

significant. There could be a lot of reasons for under-fitting of the multiple regression. At this point, the user can check the relationship between two variables. Transformation of independent variables may improve the results. This has not been demonstrated in this chapter. However, user can try out log transformation of some of the independent variables which can be used.

Table 10.4 Results of multiple regression

	Estimate	Std. error	*t*-value	*p*-value
(Intercept)	1.934	0.061	31.940	0.000
Temp range	−0.015	0.088	−0.168	0.872
Rainfall	0.237	0.115	2.063	0.085
Discharge	−0.158	0.252	−0.626	0.554
Sediment	0.018	0.158	0.116	0.912
Basin area	0.358	0.270	1.326	0.233
Latitude	−0.061	0.264	−0.232	0.824
Water temp	−0.025	0.078	−0.318	0.762

$R^2 = 0.73$; Adjusted $R^2 = 0.43$; $F_{7,6} = 2.381$; p-value $= 0.1552$

10.2.3.3 Interpretation of Regression Coefficients

It is found that none of the independent variables is found to be statistically significant (all p-value >0.05). Therefore, the present fitted multiple regression is not suitable for prediction of fish species richness.

10.2.3.4 Relative Importance of Independent Variables

Further computation is required to compute the relative importance of the variables. First, standardized estimate of each coefficient is to be computed using the following formula:

$$\widehat{b_{\text{std}}} = \frac{\widehat{b}}{\text{SE}\left(\widehat{b}\right)} \tag{10.3}$$

Note that the estimate of regression coefficient and their corresponding estimate of SE are provided in Table 10.4. These estimates are readily available in any statistical packages. Thereafter, the relative importance of each independent variable is computed as follows:

$$\text{RI}(b) = \frac{\widehat{b}^2_{\text{std}}}{\Sigma \widehat{b}^2_{\text{std}}} \tag{10.4}$$

The plot shown in Fig. 10.6 evidently suggests that sediment has the highest contribution to the fish species richness followed by Latitude > Temp range > Discharge > Water temp > Basin area > Rainfall.

10.3 Conclusion

This chapter describes application areas of multivariate analysis. Out of several multivariate analysis techniques, two frequently used techniques were discussed in a simple way without any complex mathematics involved in it. The readers, who are interested in mathematical derivation, may consult the book included in the list of

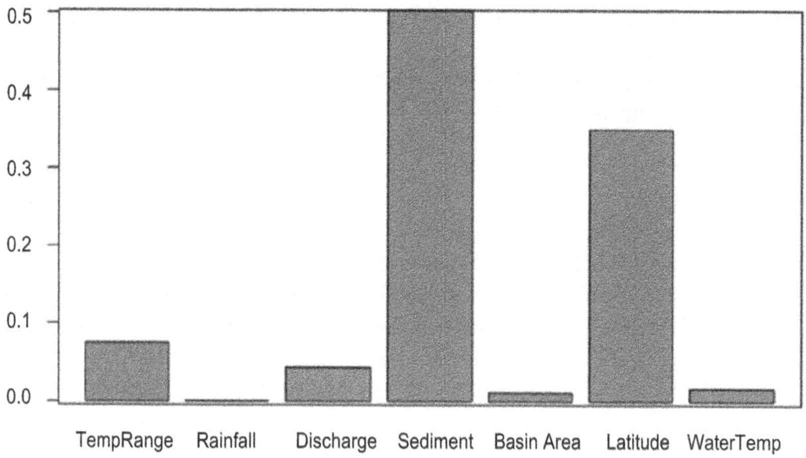

Fig. 10.6 Relative Importance of predictors of fish species richness

suggested readings. In all the cases, a generic framework of analysis has been described and demonstrated through real-life data example. However, the same can be analysed in more advanced way.

Exercise

1. The multiple linear regression analysis is a statistical framework for developing a mathematical equation that describes how
   (a) One explanatory and one or more response variables are related.
   (b) Several explanatory and several response variables response are related.
   (c) One response and one or more explanatory variables are related.
   (d) All of these are correct.
2. In regression analysis, the variable that is being predicted is the
   (a) Response, or dependent, variable
   (b) Independent variable
   (c) Intervening variable
   (d) Usually x
3. In a regression analysis, if $r^2 = 1$, then
   (a) SSE must also be equal to 1
   (b) SSE must be equal to 0
   (c) SSE can be any positive value
   (d) SSE must be negative
4. In regression analysis, the variable that is used to explain the change in the outcome of an experiment, or some natural process, is called
   (a) The independent variable or the x variable
   (b) The predictor variable

(c) The explanatory variable
(d) All are correct

5. A regression analysis between sales (in Rs. 1000) and price (in rupees) resulted in the following equation:

$$\widehat{y} = 50,000 - 8.0X$$

The above equation implies that an
(a) Increase of Rs. 1 in price is associated with a decrease of Rs. 8 in sales
(b) Increase of Rs. 8 in price is associated with an increase of Rs. 8000 in sales
(c) Increase of Rs. 1 in price is associated with a decrease of Rs. 42,000 in sales
(d) Increase of Rs. 1 in price is associated with a decrease of Rs. 8000 in sales

6. If the coefficient of determination (R^2) is a positive value, then the regression equation
(a) Must have a positive slope
(b) Must have a negative slope
(c) Could have either a positive or a negative slope
(d) Must have a positive y intercept

7. If two variables, x and y, have a very strong linear relationship, then
(a) There is evidence that x causes a change in y.
(b) There is evidence that y causes a change in x.
(c) There might not be any causal relationship between x and y.
(d) None of these alternatives is correct.

8. In regression analysis, if the independent variable is measured in kilograms, the dependent variable
(a) Must also be in kilograms
(b) Must be in some unit of weight
(c) Cannot be in kilograms
(d) Can be any units

9. In a regression analysis if SSE $=$ 200 and SSR $=$ 300, then the coefficient of determination is
(a) 0.6667
(b) 0.6000
(c) 0.4000
(d) 1.5000

10. A fitted least squares regression line
(a) May be used to predict a value of y if the corresponding x value is given.
(b) Is evidence for a cause-effect relationship between x and y.
(c) Can only be computed if a strong linear relationship exists between x and y.
(d) None of these alternatives is correct.

11. Regression analysis was applied between income (y) by sale of fish and expenditure in producing fish (x) across all wetlands of a state. The following regression function was obtained.

$$\hat{y} = 5000 + 7.25x$$

If the expenditure of two wetlands of the state differs by Rs. 30,000, then what will be the predicted difference in their sales?
(a) Rs. 217,500
(b) Rs. 222,500
(c) Rs. 5000
(d) Rs. 7.25

12. You have carried out a regression analysis; but, after thinking about the relationship between variables, you have decided you must swap the explanatory and the response variables. After refitting the regression model to the data, you expect that:
(a) The value of the correlation coefficient will change
(b) The value of SSE will change
(c) The value of the coefficient of determination will change
(d) The sign of the slope will change
13. Suppose you use regression to predict the length of a fish species by using its weight as the explanatory variable. Weight was measured in gram for a sample of 100 fish from a fish market. Now, suppose that the weight of fish is converted to kilogram. The impact of this conversion on the slope is:
(a) The sign of the slope will change
(b) The magnitude of the slope will change
(c) Both A and B are correct
(d) Neither A nor B is correct
14. Principal component analysis is used when data are
(a) Correlated
(b) Uncorrelated
(c) Both
(d) None
15. The coefficient of determination (R^2) must be
(a) Between −1 and +1
(b) Between −1 and 0
(c) Between 0 and 1
(d) Equal to SSE/(n − 2)
16. What helps in determining the optimal number of principal component in PCA analysis?
(a) Screen plot
(b) Bell curve
(c) Pie chart
(d) Box plot
17. Which principal component explains maximum variability present in original data set?
(a) Last principal component
(b) First principal component

   (c) Middle principal component
   (d) None
18. Principal component analysis can only be done if data follow?
   (a) Binomial distribution
   (b) Normal distribution
   (c) Poisson distribution
   (d) No assumption about data distribution
19. In multiple regression analysis, R^2 is also called the
   (a) Residual
   (b) Coefficient of correlation
   (c) Coefficient of determination
   (d) Standard error of the estimate
20. The main purpose(s) of multiple linear regression is
   (a) Predicting one variable on the basis of others
   (b) Explaining one variable in terms of others
   (c) Describing the relationship between one variable and others
   (d) All of the above
21. The major conceptual limitation of all regression techniques is that one can only ascertain relationships, but never be sure about underlying causal mechanism.
   (a) True
   (b) False
22. In multiple linear regression, a residual is the difference between the predicted Y and actual Y values.
   (a) False
   (b) True
23. In a PCA the thumb rule to select the number of components is
   (a) Eigenvalue < 1
   (b) Eigenvalue ≥ 1
   (c) Eigenvalue > 0
   (d) Eigenvalue < 0
24. A term used to describe the case when the independent variables in a multiple regression model are correlated is
   (a) Regression
   (b) Correlation
   (c) Multicollinearity
   (d) None of the above
25. A multiple regression model has the form: $y = 2 + 3x_1 + 4x_2$. As x_1 increases by 1 unit, y will
   (a) Increase by 3 units
   (b) Increase by 3 units holding x_2 constant
   (c) Increase by 3%
   (d) None of the above

Further Reading

Johnson RA, Wichern DW (2002) Applied multivariate statistical analysis, 5th edn. Pearson Education Inc, New Delhi

Levine MS (1977) Canonical analysis and factor comparison. Quantitative applications in the social sciences series, No. 6. Sage Publications, Thousand Oaks, CA

Romesburg HC (1984) Cluster analysis for researchers. Lifetime Learning Publications, Belmont, CA

Sharma S (1996) Applied multivariate techniques. John Wiley & Sons, New York, NY

Introduction to Electronic Spreadsheet and Microsoft Excel

11

Key Topics

- Concept of spread
- Use of an electronic spread sheet
- Data processing in Microsoft excel
- Development timelines
- Basics of Microsoft excel

Electronic spreadsheets have become essential management tools from financial management to natural resource management. In fisheries data management specifically in research, diversified data need to be processed. It may be from biological, physiochemical and geomorphological to production data. These data may be in various format; it may be string/text, quantity (numeric), date/time and spatial location (coordinate, longitude/latitude). Fisheries data are diverse in nature due to type, magnitude, units, aspects, etc. To present these data in tabular form, graphical form and report form, an electronic spreadsheet is a solution. There are various electronic spreadsheets available in the market, but here Microsoft's Excel is considered. Google Sheets is spreadsheet of Google web-based office suite and Calc of OpenOffice and Lotus 1-2-3 of SmartSuite of Lotus software.

11.1 What Is a Spreadsheet?

In the realm of accounting jargon, a 'spread sheet' or spreadsheet is a large sheet of paper with columns and rows that organizes data about transactions for a business person to examine. It spreads or shows all of the costs, income, taxes and other related data on a single sheet of paper for a manager to examine when making a decision. An electronic spreadsheet organizes information into software-defined

B. K. Das et al., *Concept Building in Fisheries Data Analysis*,
https://doi.org/10.1007/978-981-19-4411-6_11

columns and rows. The data can then be 'added up' by a formula to give a total or sum. The spreadsheet program summarizes information from many paper sources in one place and presents the information in a format to help a decision-maker see the financial 'big picture' for the company.

11.2 Spreadsheet Development Timeline

11.2.1 Beginnings and the 'Tale of VisiCalc'

In 1961, Professor Richard Mattessich, University of California at Berkeley, pioneered the development of computerized spreadsheets for use in business accounting. 'He pioneered financial spread sheet analysis and simulation and did the basic research on which such best selling micro–computer programmes as Visi–Calc, super–Calc, Lotus 1–2–3, etc. are based' (source: Spreadsheet: Its First Computerization (1961–1964) by Richard Mattessich, http://dssresources.com/history/mattessichspreadsheet.htm).

The modern era of microcomputer-based electronic spreadsheets begins with the 'Tale of VisiCalc'. The tale of VisiCalc is part myth and part fact for most of us. The story is that Dan Bricklin was preparing a spreadsheet analysis for a Harvard Business School 'case study' report and had two alternatives: (1) do it by hand or (2) use a clumsy time-sharing mainframe program. Bricklin thought there must be a better way. He wanted a program where people could visualize the spreadsheet as they created it. His metaphor was 'an electronic blackboard and electronic chalk in a classroom'.

Creators of the World's First Electronic Spreadsheet VisiCalc (The Name 'VisiCalc' Is a Compressed Form of the Phrase 'Visible Calculator')

By the fall of 1978, Bricklin had programmed the first working prototype of his concept in integer basic. The program helped users input and manipulate a matrix of 5 columns and 20 rows. The first version was not very 'powerful', so Bricklin recruited an MIT acquaintance Bob Frankston to improve and expand the program. Bricklin calls Frankston the 'co-creator' of the electronic spreadsheet. Frankston created the production code with faster speed, better arithmetic and scrolling. He also expanded the program and 'packed the code into a mere 20k of machine memory, making it both powerful and practical enough to be run on a microcomputer'. During the fall of 1978, Daniel Fylstra, founding Associate Editor of Byte Magazine, joined Bricklin and Frankston in developing VisiCalc.

11.2.2 Daniel Fylstra

Fylstra was also an MIT/HBS graduate. Fylstra was 'marketing-oriented' and suggested that the product would be viable if it could run on an Apple microcomputer. Bricklin and Frankston formed Software Arts Corporation on January 2, 1979. In May 1979, Fylstra and his firm Personal Software (later renamed VisiCorp) began marketing 'VisiCalc'. The name 'VisiCalc' is a compressed form of the phrase 'visible calculator'.

11.2.3 After VisiCalc?

The market for electronic spreadsheet software was growing rapidly in the early 1980s, and VisiCalc stakeholders were slow to respond to the introduction of the IBM PC that used an Intel computer chip. Beginning in September 1983, legal conflicts between VisiCorp and Software Arts distracted the VisiCalc developers, Bricklin and Frankston. During this period, Mitch Kapor developed Lotus, and his spreadsheet program quickly became the new industry spreadsheet standard.

11.2.4 Lotus 1-2-3

Lotus 1-2-3 made it easier to use spreadsheets, and it added integrated charting, plotting and database capabilities. Lotus 1-2-3 established spreadsheet software as a major data presentation package as well as a complex calculation tool. Lotus was also the first spreadsheet vendor to introduce naming cells, cell ranges and spreadsheet macros. Lotus 1-2-3 is still one of the all-time best-selling application software packages in the world.

11.2.5 Microsoft Excel

The next milestone was the Microsoft Excel spreadsheet. Excel was originally written for the 512K Apple Macintosh in 1984–1985. Excel was one of the first spreadsheets to use a graphical interface with pull-down menus and a point and click capability using a mouse pointing device. The Excel spreadsheet with a graphical user interface was easier for most people to use than the command line interface of PC-DOS spreadsheet products. Many people bought Apple Macintoshes so that they could use Bill Gates' Excel spreadsheet program. There is some controversy whether a graphical version of Microsoft Excel was released in a DOS version. Microsoft documents show the launch of Excel 2.0 for MS-DOS version 3.0 on October 31, 1987. When Microsoft launched the Windows operating system in 1987, Excel was one of the first application products released for it. When Windows finally gained wide acceptance with Version 3.0 in late 1989, Excel was Microsoft's flagship product.

11.3 Other Electronic Spreadsheets

11.3.1 Google Docs

Essentially, Google Docs Spreadsheet is Google's own version of an Excel-like spreadsheet application; the only differences are that it's available for free, without having to download anything on a desktop. Google Docs Spreadsheet looks pretty similar to Excel, but it actually operates inside a web browser. It has much of the same functionality as Excel including a range of charts and the ability to create formulas, so it's a pretty good alternative to curtail expenses on Microsoft Excel. What's more, Google Docs can be used anywhere, so long as there is an Internet connection. This makes working remotely on spreadsheets easier than ever, as one can be sure that files are backed up online automatically too.

11.3.2 OpenOffice

OpenOffice is a solution that is being utilized by more and more businesses these days. It's yet another extremely feature-rich and completely free alternative to Microsoft Excel. Unlike Google Docs Spreadsheet, OpenOffice is a downloadable program and operates locally on a personal computer (much like Excel). It works on Windows, Mac OSX and Linux, so operating system being used is not a hurdle. OpenOffice is pretty much guaranteed to work perfectly. In terms of functionality, OpenOffice is very similar to Microsoft Excel. It looks very similar and has charts, macros and most of the other features one might be familiar with if they have previously used.

11.3.3 Microsoft Excel

One of the benefits of OpenOffice over Google Docs Spreadsheet is that once downloaded, it will not need access to the Internet in order to use the program. It should also be noted that OpenOffice allows one to open, edit and create Excel worksheets, so it's perfectly easy to make the switch from Excel to OpenOffice without much hassle.

11.3.4 Excel Web Application

Strangely, one of the best free alternatives to Microsoft Excel is actually Excel itself. Basically, in order to compete with Google's free online services in the form of their Google Docs Apps, Microsoft launched a web version of their most popular Office applications, including Excel. Just like Google Docs Spreadsheet, Excel Web Application is available on the web free of charge. One just needs a Microsoft account to actually use it. It should be noted that this isn't the full version of Excel, and therefore, there are some features that aren't available. For example, users will not be able to edit sparklines, and VBA code is not available. However, the Excel Web Application does offer most of the functionality that most small business owners will need from a spreadsheet application, including macros and so forth.

The Excel Web Application allows one to open standard .xlsx and .xlsm file types, so once again, making the switch is simple and straightforward. But like Google Docs Spreadsheet, access to the Internet is required in order to use the application.

11.3.5 Zoho Sheet

Lastly, there is Zoho Sheet. This is another cloud-based application (like Google Docs Spreadsheet and the Excel Web Application); an Internet connection is needed in order to use it. Zoho offers a lot of cloud-based web apps, many of which are not free. All one needs to do is to create a free account with Zoho and then sign in. At this stage, the user has to simply head to 'Productivity' on the menu, and select the 'Sheet' app from the list. Zoho Sheet looks very similar to Microsoft Excel, and it allows working with local files on a computer or Google Docs files if that is a preference. Creation of files from scratch in the Zoho Sheet app can also be done. For most small business owners, this is yet another great alternative as it offers most of the functionality that are similar to Excel.

11.4 Electronic Spreadsheets and Microsoft Excel

Electronic spreadsheet is an automated version of an accountant's ledger made up of rows and columns of numerical data and replaces the calculator as tool to help analyse and evaluate numerical information.

11.4.1 Uses of an Electronic Spreadsheet

- *Data Analysis*
  A spreadsheet is an easy way to store all different kinds of data. These data types can include financial data, customer data and product data.
- *What Is Data?*
  In computing, data is information that has been translated into a form that is efficient for movement or processing. Relative to today's computers and transmission media, data is information converted into binary digital form. It is acceptable for data to be used as a singular subject or a plural subject. Excel spreadsheets can support more than a million rows and more than 16,000 columns, so one has plenty of space to store a huge amount. That's what makes them ideal for database creation.
- *Accounting Uses*
  Businesses across the world use spreadsheets to keep track of their business accounts. A user can enter formulas that will do all of their business calculations

as they go without the expense of using accounting software. Spreadsheets will do all the calculations for them, which saves them having to do those calculations manually.

- **Budgeting Help:** One can keep track of one's business spending on a spreadsheet. A custom budget can be created in spreadsheets to manage income and expenditures so that entrepreneurs do not end up with any surprises when it is time to file their tax return.
- **Data Exports:** Spreadsheets can be used to contain data that has been exported from other systems. Sometimes online data management systems do not present data in a format that is easy to read or if it is difficult to access the data offline.
- **Data Cleanup:** In spreadsheets users are able to identify inaccuracies and duplicates in data. With a spreadsheet, it is easy to get rid of unwanted or low-quality data. This is especially necessary in relation to any kind of data, which are required to be kept up to date and accurate.
- **Generating Reports and Charts:** Spreadsheets can be used to report trends or make business forecasts. These reports can be made available to everyone at the company; they can take the time to analyse the data for themselves. It is easy to make comparisons between sets of data. One can pivot tables to showcase particular aspects of data as well, which is necessary when there are huge amounts of data to sort through. Data in a spreadsheet can be used to create charts that can then be used for reporting. Graphs and pie charts can be created to condense the data in a format that is easy to read at a glance. It is a fantastic feature that can be great for business presentations that are required to showcase lots of data in a short time.
- **Business Administrative Tasks:** Spreadsheets can be used for the creation of receipts and invoices. It can also be used for business-related forecasts and planning ahead for the future.

These are some of the common purposes spreadsheets are used for, but there are additional ways like inventory management, financial planning, etc. in which a spreadsheet can be utilized to achieve business goals. They can be configured to one's suitable needs and have to be tapped in all the ways so that this simple yet effective tool can be used in different ways.

11.4.2 Advantages of Using an Electronic Spreadsheet

- Easy to move copy, edit and delete table entries
- Easy to format design and appearance of table
- Ability to perform *what-if* or *sensitivity analysis* by changing values to see how related entries are adjusted
- Easy to create charts to display relationships among values in the worksheet
- Ability to open and use multiple spreadsheets at one time

11.4.3 Anatomy of an Excel Program Window

1. Excel shares the same user interface as Word and other Microsoft Office applications. This standard interface includes the following:
   (a) Title bar
   (b) Menu bar
   (c) Standard toolbar
   (d) Formatting toolbar
   (e) Office assistant
   (f) Status bar
2. Programme window
   (a) A set of worksheets (spreadsheets)
   (b) Display the sheet on which you are currently working
   (c) An Excel document is called (.xls file extension)
3. Items in an Excel workbook
   (a) Sheet tabs.
   (b) Column headings.
   (c) Row headings.
   (d) Cells.
   (e) Active cell fill handle.
   (f) Name box.
   (g) Formula bar.
   (h) Check your Excel lab book for a figure of the window.
4. Cells: A cell is the intersection of a row and a column. For example, C4 refers to the fourth row of the *C*th column. A cell range refers to a single or a block of adjacent cells—e.g., the cell range 'A1:A5' includes the cells A1, A2, A3, A4 and A5.
5. Cell contents: A cell can contain any of the following types of information
   (a) **Value**—numbers that represent a quantity of some kind, including dates and time
   (b) **Text**—any combination of letters, symbols, numbers and spaces
   (c) **Formulas**—the arithmetic used to calculate values displayed in a worksheet. It always begins with an '=' sign and expressed as a combination of numbers, cell references, arithmetic operators, '+', '−', '*' (multiplication) and '/' (division) or function. If the value of a cell changes, Excel automatically recalculates any formula affected by the change (see *what if* analysis).
   (d) **Function**—special, pre-written formula that provides a shortcut for commonly used calculations (e.g., sum, average). Every function has a syntax, the rules specifying how the function must be typed (punctuation, parentheses, number and type of arguments). Almost every function has arguments; the values are passed to the function to be used in the calculation (usually cell references/range). Some frequently used functions are sum, average, min, max, etc.

(e) **Copying Cell Contents**—there are two methods for copying the contents of a cell: copy/paste commands and fill handle.

(f) **Formatting Cells**—the following are all different aspects of a cell that can be formatted:

- Number
- Alignment (horizontal, vertical, indent)
- Orientation
- Text control
- Character formats (font, size, colour, effects)
- Borders
- Patterns (shading)

6. Cell references

There are three types of cell references in Excel:

(a) **Relative**—cell reference in a formula that changes when copied to another cell.

Example 11.1 Cell C1 contains the formula: =A1 + B1. In this formula, both A1 and B1 are relative cell references. If C1 was copied to cell C2, the contents of C2 would be =A2 + B2 ii.

(b) **Absolute**—cell reference in a formula that does NOT change when copied to another cell.

Example 11.2 Cell C3 contains the formula: =$B$2 + B3. In this formula, $B$2 is an absolute cell reference. If C3 was copied to cell C4, the contents of C4 would be =$B$2 + B4. $B$2 remained the same because it was an absolute reference. The reference B3 changed to B4 because it is a relative reference.

(c) **Mixed**—a cell with an absolute and mixed part. For example, $B '$' in a cell reference indicates the reference to the row or column in absolute and should not change if the contents of the cell are copied.

7. Charts: Items that can be added to a worksheet and that can be sized and moved are called objects, such as textboxes, graphics, charts, etc. A chart is a visual representation of data in the spreadsheet. There are two possible methods in which a chart can be included in a workbook.

(a) **Embedded chart**—the chart object is on the same worksheet as the chart data.

(b) **Chart sheet**—the chart object is on a separate sheet with no rows and columns. Charts are usually added using the **Chart Wizard.**

11.4.3.1 Some Terminologies

1. **Data range**—data from the spreadsheet that is to be included in the chart.
2. **Nonadjacent range**—a group of individual cells or ranges of cells that are not next to each other and may even be from different worksheets, often used when specifying the data range for a chart.
3. **Data series**—a group of related data points.

11.5 Functions Related to Inland Fisheries Management

Electronic spreadsheets play a vital role in inland fisheries management just as data collected from primary and secondary sources. In the twenty-first century, there are several hurdles that need to be overcome in order to manage inland fisheries. Management has to be conceptualized at multiple scales of fish existence or their movement. This involves a lot of cooperation and interaction between organizations, scientists, researchers, planners, policy makers and administrators. This again requires large-scale fisheries data that focus on specific issues and address different objectives. The interplay of a variety of factors resulting in different fishing conditions requiring focused management strategies brings about a dynamic approach in analysing the collected data. An electronic spreadsheet enhances such calculations while making tedious manual approach to take a back seat.

11.5.1 Data Validation

It is the process of putting checks in the data entry process to efficiently avoid data entry errors. Data validation is a feature available in Microsoft Excel. It allows you to do the following:

- Make a list of the entries that restricts the values allowed in a cell.
- Create a prompt message explaining the kind of data allowed in a cell.
- Create messages that appear when incorrect data has been entered.
- Check for incorrect entries by using the **Auditing toolbar**.
- Set a range of numeric values that can be entered in a cell.
- Determine if an entry is valid based on calculation in another cell.

11.6 Data Entry by Creating Forms

1. Convert your column names into a table; go to Insert ⇑ Table.

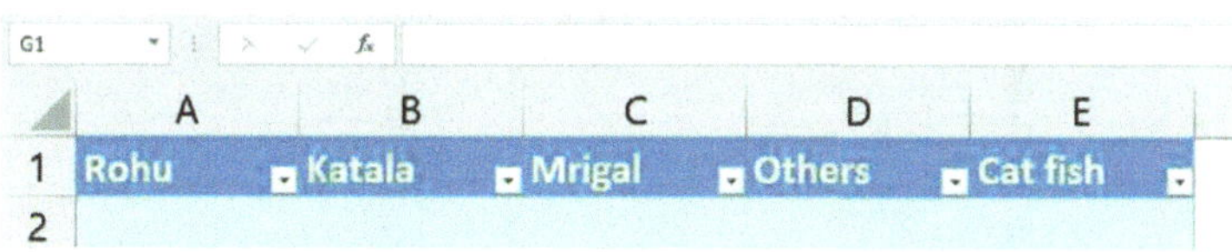

Make sure **My Table** header is checked.

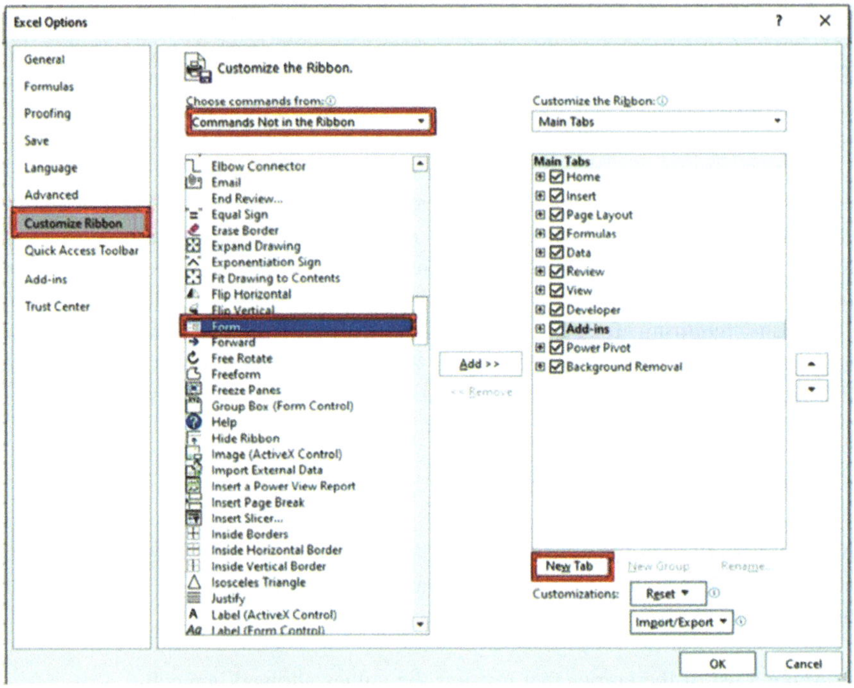
Excel Options
General
Formulas
Proofing
Save
Language
Advanced
Customize Ribbon
Quick Access Toolbar
Add-ins
Trust Center
Customize the Ribbon.
Choose commands from:
Commands Not in the Ribbon
Elbow Connector
Email
End Review...
Equal Sign
Erase Border
Expand Drawing
Exponentiation Sign
Fit Drawing to Contents
Flip Horizontal
Flip Vertical
Form...
Forward
Free Rotate
Freeform
Freeze Panes
Group Box (Form Control)
Help
Hide Ribbon
Image (ActiveX Control)
Import External Data
Insert a Power View Report
Insert Page Break
Insert Slicer...
Inside Borders
Inside Horizontal Border
Inside Vertical Border
Isosceles Triangle
Justify
Label (ActiveX Control)
Label (Form Control)
Add >>
<< Remove
Customize the Ribbon:
Main Tabs
Main Tabs
Home
Insert
Page Layout
Formulas
Data
Review
View
Developer
Add-ins
Power Pivot
Background Removal
New Tab
New Group
Rename...
Customizations:
Reset
Import/Export
OK
Cancel

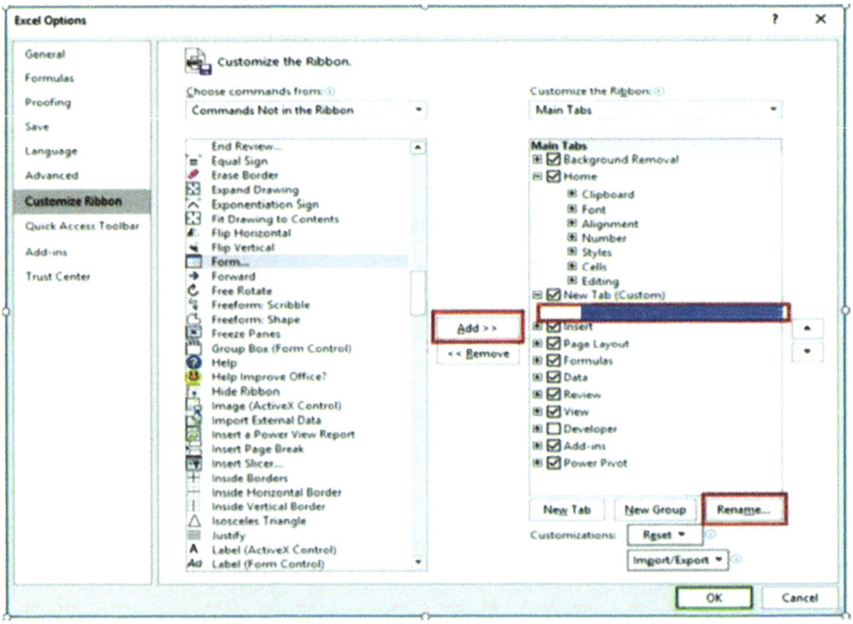
Excel Options
General
Formulas
Proofing
Save
Language
Advanced
Customize Ribbon
Quick Access Toolbar
Add-ins
Trust Center
Customize the Ribbon.
Choose commands from:
Commands Not in the Ribbon
End Review...
Equal Sign
Erase Border
Expand Drawing
Exponentiation Sign
Fit Drawing to Contents
Flip Horizontal
Flip Vertical
Form...
Forward
Free Rotate
Freeform: Scribble
Freeform: Shape
Freeze Panes
Group Box (Form Control)
Help
Help Improve Office?
Hide Ribbon
Image (ActiveX Control)
Import External Data
Insert a Power View Report
Insert Page Break
Insert Slicer...
Inside Borders
Inside Horizontal Border
Inside Vertical Border
Isosceles Triangle
Justify
Label (ActiveX Control)
Label (Form Control)
Add >>
<< Remove
Customize the Ribbon:
Main Tabs
Main Tabs
Background Removal
Home
Clipboard
Font
Alignment
Number
Styles
Cells
Editing
New Tab (Custom)
Insert
Page Layout
Formulas
Data
Review
View
Developer
Add-ins
Power Pivot
New Tab
New Group
Rename...
Customizations:
Reset
Import/Export
OK
Cancel

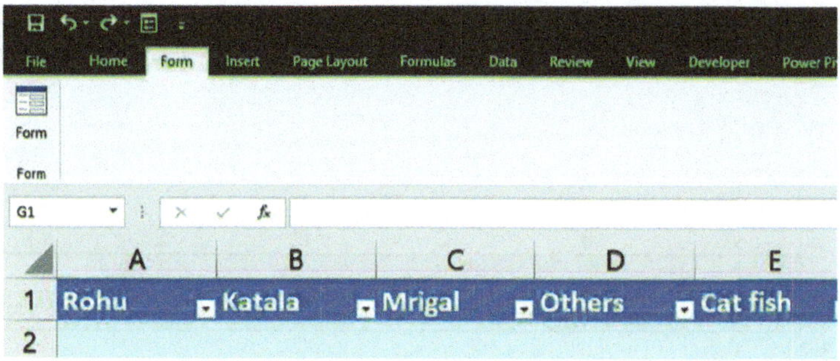

2. Let us add the Form Creation functionality. Go to File → Options.
3. Go to Customize Ribbon. Select Commands Not in the Ribbon and Form. This is the functionality we need. Click New Tab.
4. Under the **NewTab**, select **NewGroup** and click Add. This will add **Forms** to a New Tab in our Ribbon.
   Notice that there is also a **Rename** button; you can use it to rename the **New Tab** and **New Group** into something more descriptive, like Form.
5. Select your Table, and on your new **Form** tab, select **Form**.
6. A new Form dialogue box will pop up. Input your data into each section. Click **New** to save it. Repeat this process for all the records you want to add. Press Close to get out of this screen, and see the data in your Excel table. You can now use this new form to continually input data into your Excel table.

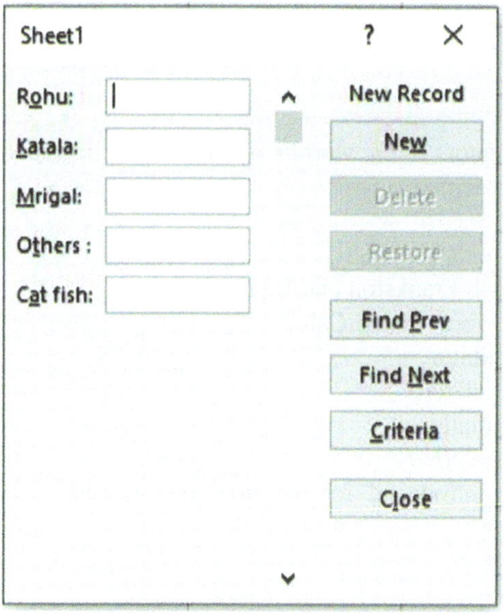

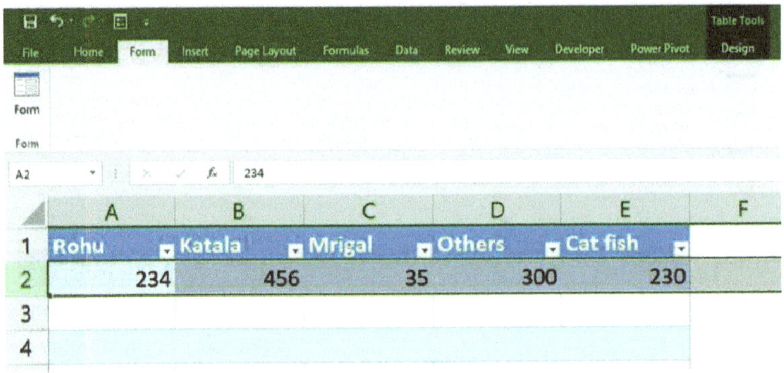

11.7 Conclusion

An electronic spreadsheet is a handy computer programme, which provides wings to the user to visualize data effectively. It provides tools for data management as well as for visualization in the form of graphs and reports. One can use from basic mathematical operation to statistical operation to analyse the data. MS Excel (electronic spreadsheet) is useful for inland fisheries database management, data analysis, graph, chart and report preparation.

Exercise

1. Who are the creators of the world's first electronic spread sheet VisiCalc?
   (a) Robert Gentleman
   (b) Steve Jobs
   (c) Bill Gates
   (d) Bricklin, Bob Frankston and Daniel Fylstra
2. What is the full form of VisiCalc?
   (a) Visual calculator
   (b) Visible calculator
   (c) Vision calculation
   (d) None of the above
3. Which vendor introduced for the first time naming cells, spreadsheet and macros?
   (a) Lotus 1-2-3
   (b) Microsoft
   (c) Apple
   (d) None of the above
4. When Microsoft Excel was introduced?
   (a) 1990
   (b) 1985

   (c) 1984
   (d) 1970
5. Which spread sheet is browser based and needs Internet connection?
   (a) Lotus 1-2-3
   (b) Zoho Sheets
   (c) Google Sheet
   (d) None of the above
6. What is the intersection of a column and a row on a worksheet called?
   (a) Value
   (b) Column
   (c) Row
   (d) Cell
7. What type of chart is useful for comparing values over categories?
   (a) Column chart
   (b) Pie chart
   (c) Line chart
   (d) None of the above
8. Which one is not a function in MS Excel?
   (a) AVG
   (b) SUM
   (c) MIN
   (d) MAX
9. Functions in MS Excel must begin with ...
   (a) No sign
   (b) A () sign
   (c) A > sign
   (d) An equal sign
10. In Excel, columns are labelled as
    (a) A, B, C, etc.
    (b) 1, 2, 3, etc.
    (c) $1, $2, $3, etc.
    (d) None of the above
11. The basic unit of a worksheet into which you enter data in Excel is called a
    (a) Table
    (b) Cell
    (c) Box
    (d) None of the above
12. How to refer the cell corresponding to column C and row 9?
    (a) col9
    (b) 9C
    (c) C9
    (d) All of the above
13. In Excel data entry, the data format is
    (a) Number
    (b) Text

   (c) Date
   (d) All of the above
14. The function SUM(1,3,6,9,11,15,33,2,7,4) will return the value
   (a) 80
   (b) 91
   (c) 95
   (d) None of the above
15. The function MAX(1,3,6,9,11,15,33,2,7,4) will return the value …
   (a) 1
   (b) 4
   (c) 33
   (d) None of the above
16. The function LEFT("ICAR-CIFRI",5) will return …
   (a) ICAR-CIFRI
   (b) -CIFRI
   (c) CIFRI
   (d) ICAR-
17. The function MEDIAN(2,4,5,7,8,9,12,3,5,8) will return …
   (a) 7
   (b) 6
   (c) 9
   (d) None of the above
18. In Excel, which one denoted a range from B1 to E5?
   (a) B1–B5
   (b) B1, B5
   (c) B1:E5
   (d) None of the above
19. An Excel file is generally called a/an:
   (a) Sheet
   (b) Spreadsheet
   (c) Worksheet
   (d) Workbook
20. In Excel what function is used to calculate mean of a series?
   (a) Mean
   (b) Average
   (c) STD
   (d) None of the above

Further Reading

Fischer W (2016) Excel: quick start guide - from beginner to expert. https://www.microsoft.com/en-in/microsoft-365/excel, https://support.microsoft.com/en-us/office/excel-video-training

Introduction to R Software 12

Key Topics

- To get acquaintance with R
- Arithmetic operations
- Handling data frames
- Getting started
- Data types
- Importing data files

R is a free software environment for statistical computing and graphics. It is a command line-driven program and runs on a wide variety of UNIX platforms, Windows and Mac OS. It is an interpreter, i.e., the user enters commands at the prompt, and each command is executed one at a time. R is very much a vehicle for newly developing methods of interactive data analysis. It has developed rapidly and has been extended by a large collection of packages. Over the last 10 years, R has become one of the most widely used statistical software packages among statisticians and researchers since it provides more than 4000 specialized packages.

A Brief History John Chambers and colleagues developed R at Bell Laboratories. R is an implementation of the S programming language and combines with lexical scoping semantics inspired by Scheme. R was named partly after the first names of two R authors. The project conceives in 1992, with an initial version released in 1995 and a stable beta version in 2000.

B. K. Das et al., *Concept Building in Fisheries Data Analysis*,
https://doi.org/10.1007/978-981-19-4411-6_12

12.1 Why Use R?

1. **Data Manipulation:** R allows the data scientist to shape the data-set into a format that could be easily accessed and analysed by slicing large multivariate datasets. In my opinion, it is one of the very few tools that have great indexing techniques.
2. **Data Analysis:** Any kind of statistical data analysis could be found in R. R is an open source development tool that is supported by a large community of statisticians and computer scientists. It has over 4000 packages that implement various statistical analysis tools related to hypothesis testing, model fitting, clustering techniques and machine learning.
3. **Data Visualization:** R is 'the package' for data visualization. There are many on the shelf graph functions in R that are ready to be used. The best part of R is that it gives the developer capabilities to implement any visualization idea for any data set. In addition, animated and interactive graphs could be implemented easily in R.

12.1.1 Programming Features of R

1. R supports procedural programming with functions and object-oriented programming with generic functions. Procedural programming includes procedure, records, modules, and procedure calls, while object-oriented programming language includes classes, objects and functions.
2. R is an interpreted language. So we can access it through command line interpreter.
3. R's programming features include database input, exporting data, viewing data, variable labels, missing data, etc. It has effective data handling and storage facilities.
4. R supports matrix arithmetic, and it supports a large pool of operators for performing operations on arrays and matrices.
5. R has the most advanced graphical capabilities, as well as facilities to print the reports for the analysis performed in the form of graphs, either on-screen or on hard copy.

12.2 Getting Started

12.2.1 Installation (for Windows Users)

12.2.1.1 To Install R

1. Open an Internet browser and go to www.r-project.org
2. Click the 'download R' link in the middle of the page under 'Getting Started'.
3. Select a CRAN location (a mirror site) and click the corresponding link.
4. Click on the 'Download R for Windows' link at the top of the page.

5. Click on the 'install R for the first time' link at the top of the page.
6. Click 'Download R for Windows', and save the executable file somewhere on your computer. Run the .exe file and follow the installation instructions.

Now that R is installed, you need to download and install RStudio.

12.2.1.2 To Install RStudio

1. Go to www.rstudio.comand; click on the 'Download RStudio' button.
2. Click on 'Download RStudio Desktop'.
3. Click on the version recommended for your system, or the latest Windows version, and save the executable file.
4. Run the .exe file and follow the installation instructions.

12.2.1.3 To Install the SDSFoundations Package

1. Download SDSFoundations to your desktop (make sure it has the '.zip' extension).
2. Open RStudio. Click on the Packages tab in the bottom right window. Click 'Install'.
3. Select install from 'Package Archive File'.
4. Select the SDSFoundations package file from your desktop.
5. Click install. You are done! You can now delete the SDSpackage file from your desktop.

12.2.2 Working with R Scripts

R provides the freedom of selecting and editing tools to interact with the native console. While scripting in R, you don't need to type commands but rather call functions to achieve results. The RConsole allows command editing.

The developed R scripts can be executed in the selected editor. You could download the data and save them in a local file, or just cut and paste the data from the browser to an editor such as Notepad, and then save them. The prominent editors available for R programming language are (1) **RGui** (R graphical user interface) and (2) **RStudio**.

12.2.3 R Graphical User Interface (RGui)

Once you download R, RGui is provided as the standard graphical user interface (GUI). The most important component of RGui is the R console window. The console window in R is a place where instructions, scripts and general R operations are performed. The console window also has R tools to manage the R environment. The R console screen appears every time the RGui is opened. It lists some basic information such as the R version installed and the licensing conditions.

In the RGui window, you can open a new script, go to the 'File' menu and select 'New Script'. The RGui can be accessed using the menu shortcuts created during the installation process.

In the R prompt, a '>' symbol indicates the place where the user can enter commands. To quit an active R session, you need to type the following code in the console after the command prompt (>):

```
> q()
```

R asks a question to ensure that the user wishes to quit the active session.

Note the parentheses after the q; this is because in R you don't type commands but rather call functions to achieve results and even quit.

12.2.4 RStudio

RStudio is an integrated development environment (IDE) for R language. RStudio is a code editor and development environment, with some nice features that make code development in R easy and fun.

1. Features of R Studio
   (a) Code highlighting that gives different colours to keywords and variables, making it easier to read.
   (b) Automatic bracket matching.
   (c) Code completion, so as to reduce the effort of typing the commands in full.
   (d) Easy access to R Help, with additional features for exploring functions and parameters of functions.
   (e) Easy exploration of variables and values. RStudio is available free of charge for Linux, Windows and Mac devices. It can be directly accessed by clicking the RStudio icon in the menu system on the desktop.
2. Components of R Studio
   (a) **Source:** Top left corner of the screen contains a text editor that lets the user work with source script files. Multiple lines of code can also be entered here. Users can save the R script files to the disk and perform other tasks on the script.
   (b) **Console;** On bottom left corner is the R console window. The console in RStudio is identical to the console in RGui. All the interactive work of R programming is performed in this window.
   (c) **Workspace and History:** On the top right hand corner is the R workspace and the history window. This provides an overview of the workspace, where the variables created in the session along with their values can be inspected. This is also the area where the user can see a history of the commands issued in R.

3. The **Files, Plots, Package** and **Help** buttons on the bottom right-hand corner give access to the following tools:
   (a) **Files:** This is where the user can browse folders and files on a computer.
   (b) **Plots:** This is where R displays the user's plots.
   (c) **Packages:** This is where the user can view a list of all the installed packages.
   (d) **Help:** This is where one can browse the built-in help system of R.

12.2.5 R Scripting

Now, let us finally start scripting in R. Let's create a script to print **Hello world!** in R. To create scripts in R, you need to perform the following steps:

Here in R, you will have to enclose some commands in print() to get the same output as on the command line. So you need to type the command given below. This takes *Hello World* as the input in R.

```
>print("Hello#$World!")
```

You should get the following as the output on the console:

```
[1] "Hello#$World!"
```

Congratulations! You have successfully executed your first command in R.

12.2.6 Sourcing a Script in R

R provides the feature to develop scripts in external editors and then import in R for execution. This is known as sourcing a script. The individual line of code or a block or entire script can be sourced into R on RGUI or RStudio as shown below.

1. To send an individual line of code from the editor to the console: In RGUI, click the line of code, and then press **Ctrl + R** to execute the instruction.
   In RStudio, press **Ctrl + Enter** or **Click the Run button** to execute the instruction.
2. To send a block of highlighted code to the console: Select the block that needs to be executed, and then press **Ctrl + R** in RGui or **Ctrl + Enter** in RStudio.
3. To send the entire script to the console:
   In RGui, click anywhere in the script window, and then choose **Edit ↑ Run all.**
   In RStudio, click anywhere in the source editor, and press **Ctrl + Shift + Enter**, or click the **Source.**

These keyboard shortcuts are defined only in RStudio. In different source editors, these options may be different.

Now the entire script can be sent to the R console. To do this, click the **Source button** at the top right-hand corner of the editor window, or choose **Edit ↑ Source**. The script reaches the point where it asks for input and then waits for the user to enter the data into the console window.

12.3 Arithmetic and Logical Operators

12.3.1 Arithmetic Operators

As most of the programming languages, R can perform basic arithmetic operations. There is a certain number of operations that R can do, such as addition (+), subtraction (−), multiplication (*), exponentiation (^ or **), division (/) and modulo (%%). Here are some examples which you can try yourself.

Remember, R is an interpreted language, so commands are executed one at a time.

```
> 70+5
[1] 75
> 70-5
[1] 65
> 70*5
[1] 350
> 70/5
[1] 14
> 5^4
[1] 625
> 5**4
[1] 625
> 70%%5
[1] 0
```

12.3.2 Comments

If you want to leave a comment, use a hash (#); leaving comments is crucial, because it will help you or someone else to understand your code better.

Comments are ignored by the compiler. In fact, they are treated as white space.

```
>#Let's leave a comment!
> 3+3 #Another comment
[1] 6
```

As observed above, the comments are ignored even after the arithmetic operation.

12.3.3 Logical Operators

After considering arithmetic operators, we look at the logical operators that are there: exact equality (==), exact inequality (!=), less than (<), less than or equal to (<=), greater than (>) and greater than or equal to (>=). In ordinary arithmetic, these logical operators have the value either true (1) or false (0), as shown below.

```
> 68 != 68
[1] FALSE
> 68 == 68
[1] TRUE
> 1 <= 2
[1] TRUE
> 7 >= 4
[1] TRUE
> 8 < 3
[1] FALSE
```

12.4 Data Types and Objects in R

Data are the most basic ingredients used in 'data analysis'. R supports a wide variety of data types including scalars, vectors, matrices, data frames and lists. We will go over some commonly used data types and briefly cover the idea of 'Object' in the end.

12.4.1 Scalars

In computer programming, scalar refers to an atomic quantity that can hold only one value at a time. Scalars are the most basic data types that can be used to construct more complex ones. Let's take a look of some common types of scalars with simple R commands.

12.4.1.1 Number

```
> x <- 1
> y <- 2.5
>class(x)
[1] "numeric" >class(y)
[1] "numeric"
>class(x+y)
[1] "numeric"
```

12.4.1.2 Logical Value

```
> m <- x > y    # Is x larger than y?
> n <- x < y    # Is x smaller than y?
> m
[1] FALSE
> n
[1] TRUE
>class(m)
[1] "logical"
>class(NA) # NA is another logical value: 'Not Available'/Miss-
ing Values
[1] "logical"
```

Here are some logical operations you may want to try.

```
> m &n          # AND
[1] FALSE
> m | n         # OR
[1] TRUE
>!m[1] TRUE     # Negation
```

12.4.1.3 Characters and Strings

```
> a <- "1"; b <- "2.5"    # Are they different from x and y we
used earlier?
> a;b
[1] "1"
[1] "2.5"
```

(continued)

```
> a+b                         # a+b=3.5?
Error in a + b : non-numeric argument to binary operator
>class(a)
[1] "character"
>class(as.numeric(a))       # but you can coerce this character
into a number
[1] "numeric"
>class(as.character(x)) [1] "character"    # vice versa
```

12.4.2 Vectors

A vector is a sequence of data elements of the same basic type.

```
> o <- c(1,2,5.3,6,-2,4).                       # Numeric vector
> p <- c("one","two","three",
   "four","five","six")                         # Character vector
>q <- c (TRUE,TRUE,FALSE,TRUE,FALSE,TRUE)        # Logical vector
> o;p;q
[1] 1.0 2.0 5.3 6.0 -2.0 4.0
[1] "one" "two" "three" "four" "five" "six"
[1] TRUE TRUE FALSE TRUE FALSE
```

You can extract the components of a vector in the following way:

```
> o[q]     # Logical vector can be used to extract vector
   components
[1] 1 2 6 4
>names(o) <- p    # Give each component a name
> o
one     two     three     four     five     six
1.0     2.0     5.3       6.0      -2.0     4.0
> o["three"]        # Extract your components by "calling" their
names
three 5.3
```

12.4.3 Matrix

A matrix is a collection of data elements arranged in a two-dimensional rectangular layout. Same as vector, the components in a matrix must be of the same basic type.

The following is an example of a matrix with four rows and three columns.

```
>t <- matrix(
+     1:12,                          # the data components (Don't
type "+"!)
+     nrow=4,                        # number of rows
+     ncol=3,                        # number of columns
+     byrow = FALSE)                 # fill matrix by columns
>t      [,1]    [,2].    [,3]        # print the matrix
[1,]      1       5        9
[2,]      2       6       10
[3,]      3       7       11
[4,]      4       8       12
```

Similar to vectors, matrices also use [ ] to reference elements.

```
>t[2,3]                  #component at 2nd row and 3rd column
[1] 10
>t[,3]                   #3rd column of matrix
[1] 9 10 11 12
>t[4,]                   #4th row of matrix
[1] 4 8 12
>t[2:4,1:3]
[;1] [,2] [,3]
[1,]      2    6     10
[2,]      3    7     11
[1,]      4    8     12
```

12.4.4 Data Frames

A data frame is more general than a matrix, in that different columns can have different basic data types. Data frame is the most common data type we are going to use in this section.

Extracting components from data frames is somehow similar to what we did for matrices, but after assigning names to each column (variable), it becomes more flexible.

```
> d <- c(1,2,3,4)
> e <- c("red", "white", "red", NA)
> f <- c(TRUE,TRUE,TRUE,FALSE)
> mydata <- data.frame(d,e,f)
>names(mydata) <-c("ID","Color","Passed")  # variable names
> mydata
ID Color Passed
1  1 red TRUE
2  2 white TRUE
3  3 red TRUE
4  4 <NA> FALSE
> mydata$ID          # try mydata["ID"] or mydata[1]
[1] 1 2 3 4
> mydata$ID[3]       # try mydata[3,"ID"] or mydata[3,1]
[1] 3
> mydata[1:2,]       # first two records
ID Color Passed
1   1    red    TRUE
2   2    white TRUE
```

12.4.5 Lists

A list is a generic vector containing other objects. There is no restriction on data types or length of the components. Usually, we work with lists that have named components.

```
> l <-list(vec=p, mat=t, fra=mydata, count=3) # a list with a
vector, a matrix, a data frame defined earlier and a scalar
> l
$vec
[1] "one" "two" "three" "four" "five" "six"
$mat
[,1]     [,2]     [,3]
[1,]      1        5     9
[2,]      2        6     10
[3,]      3        7     11
[4,]      4        8     12
$fra
ID Color Passed
1  1 red TRUE
2  2 white TRUE
```

(continued)

```
3  3 red TRUE
4  4 <NA> FALSE
$count
[1] 3
> l$vec                          #extract components from list
[1] "one" "two" "three" "four" "five" "six"
> l$mat [2,3]
[1] 10
> l$fra$Color
[1] red white red <NA>
Levels: red white
```

12.4.6 Objects

In R, all types of data are treated as objects. However, objects are not simply collections of data. They are particular instances of particular classes. Operations, or functions, are defined for specific classes. Let's try working on something such as a point pattern.

```
# This time I will not show R outputs with codes. Just type or
paste these lines into
R and see what you get.
x <- rnorm(50, 10, 3)    # creates 50 random x values from a
normal distribution
y <- rnorm(50, 10, 4)    # creates 50 random y values
mypoints <- as.data.frame(cbind(x,y))    # makes a data frame
class(mypoints)
mypoints summary(mypoints)
plot(mypoints)    # Gee, it looks like a point
pattern...
box <- bbox(mypoints) # Type in library(splancs) first.Bounding
Box - did this work?
Why not?
```

It seems that most functions above work well with this data frame but, 'bbox' does not. See help(bbox). It didn't work because "bbox" doesn't work on objects of class data.frame. 'bbox' operates on objects of class points (or a matrix of x and y values). Therefore you need to change the class accordingly. The following four approaches all work (try each one separately):

```
box <- bbox(cbind(x,y))
box <- bbox(as.matrix(mypoints))
box <- bbox(as.points(x,y))
box <- bbox(as.points(mypoints))
```

12.5 Handling Data Frames

A data frame is a table or a two-dimensional array-like structure in which each column contains values of one variable and each row contains one set of values from each column.

The following are the characteristics of a data frame.

- The column names should be non-empty.
- The row names should be unique.
- The data stored in a data frame can be of numeric, factor or character type.
- Each column should contain same number of data items.

12.5.1 Creating a Data Frame

The following is an example on how to create a data frame:

```
# Create the data frame.
>species.data<-data.frame(site_id=c(1:5), # setting id to site
site_name=c
("Farakka","Nabadwaip","Bolagarh","Bally","Godakhali"),
    # Name of the sites
sample_date=as.Date(c("2012-01-01","2012-02-01", "2012-03-01",
    "2012-04-01", "2012-05-01")), speciesno=c(54,21,34,45,44) #
    Species richness,
#Print the created data frame
>print(species.data) or species.data
```

When we execute the above code, it produces the following result:

```
site_id         site. _name    sample_date    species    no
1          1    Farakka        2012-01-01     54
2          2    Nabadwaip      2012-02-01     21
3          3    Bolagarh       2012-03-01     34
4          4    Bally          2012-04-01     45
5          5    Godakhali.     2012-05-01     44
```

The top line of the table, called the header, contains the column names. Each horizontal line afterwards denotes a data row, which begins with the name of the row and then followed by the actual data. Each data member of a row is called a cell.

To retrieve data in a cell, we would enter its row and column coordinates in the single square bracket '[ ]' operator. The two coordinates are separated by a comma. In other words, the coordinates begin with row position, then followed by a comma, and end with the column position. The order is important.

12.5.2 Summary of Data

The statistical summary and nature of the data can be obtained by applying ***summary*** *()* function.

```
# Create the data frame.
emp.data <- data.frame(
emp_id = c (1:5),
emp_name = c("Rick","Dan","Michelle","Ryan","Gary"),
salary = c(623.3,515.2,611.0,729.0,843.25),
start_date = as.Date(c("2012-01-01", "2013-09-23", "2014-11-15",
"2014-05-11",
"2015-03-27")),
    stringsAsFactors = FALSE
)
# Print the summary.
print(summary(emp.data))
```

When we execute the above code, it produces the following result:

```
>summary(species.data)
site_id     site_name        sample_date          speciesno
Min. :1     Bally :1         Min. :2012-01-01     Min. :21.0
1st Qu.:2.  Bolagarh :1      1st Qu.:2012-02-01   1st Qu.:34.0
Median :3   Farakka :1       Median :2012-03-01   Median :44.0
Mean :3     Godakhali:1      Mean :2012-03-01     Mean :39.6
3rd Qu.:4   Nabadwaip:1      3rd Qu.:2012-04-01   3rd Qu.:45.0
Max. :5     Max.:2012-05-01  Max. :54.0
```

12.5.3 Extracting Data from a Data Frame

Extracting specific column from a data frame using column name.

```
# Create the data frame.
>#### Creating new data frame by Extracting specific columns
>results=data.frame(species.data$site_name,species.data$speciesno)
>print(results)
```

When we execute the above code, it produces the following result:

```
species.data.    site_name species.data.    speciesno
1                Farakka                    54
2                Nabadwaip                  21
3                Bolagrah                   34
4                Bally                      45
5                Godakhali                  44
```

Extracting the predefined rows and then all columns. Now output will not be shown. Just try these codes.

```
>dat1=species.data[1:2,] #### Select first two rows and save
the newdata frame to dat1
>dat2= species.data[c(1,5),] #### Select row 1 and 5, and save
the data in dat2
```

12.5.4 Expanding the Data Frame

A data frame can be expanded by adding columns and rows.

12.5.4.1 Adding Columns

We need to just add the column vector using a new column name, as shown below:

```
###### Note length of vector should be same as the dataframe
>species.data$catch=c(500,300,600,800,1000)
>species.data ##### See whether the column had been added or no
print(species.data)
```

12.5.4.2 Adding Rows

To add more rows permanently to an existing data frame, we need to bring in the new rows in the same structure as the existing data frame and use the ***rbind()*** function.

In the example below, we create a data frame with new rows and merge it with the existing data frame to create the final data frame.

```
>new.data=data.frame(site_id=c(6:7)
site_name=c("Diamond","Namkhana"),
sample_date=as.Date(c("2012-06-01","2012-07-01")),
speciesno=65,
catch=1300
)
>species.dat2=rbind(species.data,new.data)
```

12.6 Importing Data from Files

It is often necessary to import sample textbook data into R before you start working on your homework. This section covers the ways in which you can import your stored data into R.

12.6.1 Working Directory

The data you wish to import should be in your R working directory. To find the current working directory, you can use the getwd function.

```
>getwd()    # get current working directory
```

You can select a different working directory which contains your files and data with the function setwd() and thus avoid entering the full path of the data files.

```
>setwd("<new#$path>") # set working directory
```

Note that the forward slash should be used as the path separator even on Windows platform.

```
>setwd("C:/MyDoc")
```

Once you are in your desired working directory, you can start importing different types of files containing your data.

12.6.2 Importing from Excel

Quite frequently, the sample data is in Excel format and needs to be imported into R prior to use. For this, we can use the function read.xls from the gdata package. It reads from an Excel spreadsheet and returns a data frame. The following shows how to load an Excel spreadsheet named 'mydata.xls'. This method requires Perl runtime to be present in the system.

```
>library(gdata)                      # load gdata package
>help(read.xls)                      # documentation
> mydata = read.xls("mydata.xls")    # read from first sheet
```

Alternatively, we can use the function loadWorkbook from the XLConnect package to read the entire workbook and then load the worksheets with readWorksheet. The XLConnect package requires Java to be pre-installed.

```
>library(XLConnect)     # load XLConnect package
> wk = loadWorkbook("mydata.xls")
>df = readWorksheet(wk, sheet="Sheet1")
```

12.6.3 Importing from Table File

```
100    a1    b1
200    a2    b2
300    a3    b3
400    a4    b4
```

A data table can reside in a text file. The cells inside the table are separated by blank characters. Here is an example of a table with four rows and three columns.

You can copy and paste the table above in a file named 'mydata.txt' with a text editor. Then load the data into the workspace with the function read.table.

12.6.4 Importing from CSV Files

The sample data can also be in comma separated values (CSV) format. Each cell inside such data file is separated by a special character, which usually is a comma, although other characters can be used as well. The first row of the data file should contain the column names instead of the actual data. We can read the data using the ***read.csv()*** function. Suppose we have a 'demo-data.csv' file saved in the working directory. We can import the file and save into a data frame by using the following command:

```
> mydata = read.table("mydata.txt")      # read text file
> mydata     # print data frame
           V1     V2     V3
1          100    a1     b1
2          200    a2     b2
3          300    a3     b3
4          400    a4     b4

dat=read.csv("demo-data.csv",header=T)
```

When full path is unknown to the users, the following command may be used to select the .csv file by clicking the mouse.

```
dat=read.csv(file.chose(),header=T)
```

In various European locales, as the comma character serves as the decimal point, the function read.csv2 should be used instead. For further detail of the read.csv and read.csv2 functions, please consult the R documentation.

Note *In the latest release, R studio offers GUI-based import facility directly from Excel and comma separated files. Make sure that* ***readxl*** *and* ***readr*** *packages are installed.*

Once the data file is imported and saved in a data frame (e.g. 'dat' in the above example), all the operations of data frame can be implemented. For example, first few rows can be seen by issuing command.

```
#### to view first few rows of in the data frame named as 'dat'
>head(dat)
```

R is a function-based interpreter. Every function is called from some library. During installation the base packages are installed. It becomes necessary to install extra packages for enhancing extra functionality in R. It is important to learn how to install extra packages in R and load into R. There are mainly two methods of installing extra packages as follows:

Method 1: Command based—The following command is useful for installing a known package. For example, we want to install ***ggplot*** package which enhances graphics capability.

```
#### Make sure internet is on then issue the following
command
>install.packages("ggplot")
```

Method 2: Using RStudio graphical interface—In the Menu simply click on the **Tools ↑ install packages**, and follow the on-screen instructions. After installation same function ***library()*** is used to load the package.

12.7 Conclusions

This chapter describes only some of the basics of R. The capability of R is huge. It is the complete statistical computing environment. This chapter is provided with the preliminary guidance. Anyone with logical mindset can do anything in R. Though the learning curve of R is very steep, it is extremely useful once habituated. Some guidelines are provided here to learn R.

1. Use built-in help systems with examples in R.
2. Use Google search with a very specific keywords of your need. In 99% of cases, you will get solution with sample code.
3. Watch instructive videos.
4. Read blogs, as R has huge community, and there are people who can give you support.
5. Read books and online resources. The following books are for the beginners.
   (a) Learning R: A step-by-step function guide to data analysis by Richard Cotton
   (b) R in a Nutshell: A desktop quick reference by Joseph Adler

Exercise

1. The primary R system is available from the —.
   (a) CRAN
   (b) CRWO
   (c) GNU
   (d) CRDO
2. Which of the following is used for statistical analysis user interface in R language?
   (a) Studio
   (b) RStudio
   (c) Heck
   (d) Heck
3. What is the output of the R command **print("Hello World")**?
   (a) Hello World
   (b) Hello World!
   (c) "Hello World"
   (d) Hello World!
4. Which of the following commands is used to quit R user environment?
   (a) quit
   (b) q:
   (c) q()
   (d) exit
5. If you execute the following command, what will be the result?

```
>3^3
```

   (a) 9
   (b) 27
   (c) 81
   (d) 6
6. What is the value obtained by executing the following command?

```
>3^3 + 2*2 - 6/3*3
```

   (a) 12
   (b) 25
   (c) 30
   (d) 21

7. What is the result for the following command?

```
> 10 !=2
```

   (a) True
   (b) False
   (c) 10
   (d) 20
8. Which of the command is used to store the vector values 5, 4, 3, 2 and 9 to a variable, say, *x*?
   (a) x<-5,4,3,2,9
   (b) x<-5:4:3:2:9
   (c) x<-c(5,4,3,2,9)
   (d) None of the above
9. What will be the output of the following R commands?

```
>x<-1:10
>x
```

   (a) 1 2 3 4 5 6 7 8 9 10
   (b) 1:10
   (c) 11
   (d) 1 10
10. The species richness at 10 sites are 45, 28, 12, 9, 16, 7, 23, 11, 12 and 70 in a river stretch. Which is the correct R command to compute mean species richness of the river stretch?
   (a)

```
x<-c(45,18,12,9,16,7,23,11,12,70)
m(x)
```

   (b)

```
x<-c(45,18,12,9,16,7,23,11,12,70)
mean(x)
```

   (c)

```
x<-c(45,18,12,9,16,7,23,11,12,70)
mean x
```

   (d) None of the above

11. For the species richness data in Question No. 10, specify the correct R command to compute standard deviation of the species richness.
    (a)

```
x<-c(45,18,12,9,16,7,23,11,12,70)
s(x)
```

    (b)

```
x<-c(45,18,12,9,16,7,23,11,12,70)
SD. (x)
```

    (c)

```
x<-c(45,18,12,9,16,7,23,11,12,70)
SD x
```

    (d)

```
x<-c(45,18,12,9,16,7,23,11,12,70)
sd (x)
```

12. For the species richness data in Question No. 10, what will be the output of the following command?

```
x<-c(45,18,12,9,16,7,23,11,12,70)
min(x)
```

    (a) 70
    (b) 12
    (c) 7
    (d) None of the above
13. For the species richness data in Question No. 10, what will be the output of the following command?

```
x<-c(45,18,12,9,16,7,23,11,12,70)
   max(x)
```

   (a) 70
   (b) 12
   (c) 7
   (d) None of the above
14. For the species richness data in Question No. 10, what will be the output of the following command?

```
x<-c(45,18,12,9,16,7,23,11,12,70)
    range(x)
```

   (a) None of the above
   (b) 63
   (c) 70 7
   (d) 7 70
15. For the species richness data in Question No. 10, which of the following R commands is correct to compute variance of the species richness?
   (a)

```
x<-c(45,18,12,9,16,7,23,11,12,70)
Variance(x)
```

   (b)

```
x<-c(45,18,12,9,16,7,23,11,12,70)
variance(x)
```

   (c)

```
x<-c(45,18,12,9,16,7,23,11,12,70)
Var(x)
```

   (d)

```
x<-c(45,18,12,9,16,7,23,11,12,70)
var(x)
```

16. Which of the following statements is true for a matrix in R?
   (a) Each column length is different.
   (b) The number of rows and the number of columns are equal.

(c) Elements are of the same data type.
(d) Elements are numeric only.

17. Which of the following is not true for a data frame in R?
(a) The number of rows and the number of columns are equal.
(b) Column can be of numeric, factor or character data type.
(c) Length of each columns is equal.
(d) Every column has a name.

18. Suppose the following data set is saved in a csv file. The file name is 'fish.csv'.

```
Site    Species    DO
S1      25         8
S2      50         7
S3      40         8.1
S4      35         7.5
S5      20         6
```

How to read this data into R and store it into a variable, say, dat?

(a) `dat<-read.csv (fish.csv,header=TRUE)`

(b) None of the above

(c) `dat<-read(file="fish.csv")`

(d) `dat<-read.csv (file="fish.csv",header=TRUE)`

19. Suppose a researcher wish to add a new column of catch data (in kg) series of 500, 300, 600, 800 and 1000 in the data set in Question No. 18. Which is the correct R command to add the column?
(a)

```
dat<-read.csv (file="fish.csv",header=TRUE)
dat$Catch<-c(500,300,600,800,1000)
```

(b)

```
dat<-read.csv (file="fish.csv",header=TRUE)
dat<-addcolumn(dat,c(500,300,600,800,1000))
```

(c)

```
dat<-read.csv (file="fish.csv",header=TRUE)
dat<-rbind(dat,c(500,300,600,800,1000))
```

(d) None of the above

20. For the example data set given in Question No. 18, which command is used for descriptive statistics of all the columns?

(a)

```
dat<-read.csv (file="fish.csv",header=TRUE)
ds(dat)
```

(b)

```
dat<-read.csv (file="fish.csv",header=TRUE)
descriptive(dat)
```

(c)

```
dat<-read.csv (file="fish.csv",header=TRUE)
dessatat(dat)
```

(d)

```
dat<-read.csv (file="fish.csv",header=TRUE)
summary(dat)
```

Further Reading

Dalgaard P (2002) Introductory statistics with R. Springer, New York, NY ISBN 0-387-95475-9. http://www.biostat.ku.dk/~pd/ISwR.html

Fox J (2002) An R and S-plus companion to applied regression. Sage Publications, Philadelphia, PA ISBN 0-761-92280-6 (softcover) or 0-761-92279-2. http://socserv.socsci.mcmaster.ca/jfox/Books/Companion/

Iacusand SM, Masarotto G (2002) Laboratorio di statistica con R. McGraw-Hill, Milan ISBN 88-386-6084-0 (in Italian). http://www.ateneonline.it/LibroAteneo.asp?item_id=1436

Introduction to Graphics in R

13

Key Topics
- General features of R graphics
- Standard graphics
- Customizing graphs
- Statistical graphics

Graphics is extremely useful for data exploration, visualization, analysis and presentation. Presentation of result of statistical analysis is essential for good report writing. Graphical presentation of results is more useful and attractive than tabular presentation or other. A good statistical analysis might not be appealing without graphical presentation.

Graphical facilities are an important and extremely versatile component of the R environment. It is possible to use the facilities to display a wide variety of statistical graphs and also to build entirely new types of graph. In fact R-Graphics can do anything that anyone can think of or visualize. There are several types of graphics in R, including standard graphs, maps, interactive graphics, etc. Several applications for specific R packages are also available, which provide attractive graphics. R includes at least three graphical systems, the **standard graphics** package, the **lattice** package for **Trellis** graphs and the grammar-of-graphics **ggplot2** package. In this chapter, we will describe and demonstrate some basics of graphics using standard graphics in R.

13.1 Standard Graphics in R

Before getting started with graphics in R, let us look at the basic plots that can be produced in R using the ***demo()*** function.

B. K. Das et al., *Concept Building in Fisheries Data Analysis*,
https://doi.org/10.1007/978-981-19-4411-6_13

```
>demo(graphics)
```

The above command will display the graph and associated 'r code'.

In this section we present what you need to know if you want to customize your graphs in the default graph system. Graphics functions are mainly one of two types:

1. High-level functions—complete plots, e.g. **plot(), boxplot(), dotplot()**, etc.
2. Low-level functions—add to an existing plot, such as **lines(), points(), legend(), arrows(), polygon(), text()**, etc.

Some functions can work either way, via an argument add=TRUE/FALSE, e.g. **symbols**(x, y, . . ., add=TRUE).

- The **plot()** is the most important function in traditional graphics. It is designed as a generic function that does different things with numeric data (x, y), factors (FAC), matrices (MAT), etc. The arguments can be a single point such as 0 or c(.3,.7), a single vector, a pair of vectors or many other R objects.
- par() is another important function which defines the default settings for plots.
- There are many other plot functions which are specific to some tasks such as hist (), boxplot(), etc. Most of them take the same arguments as the plot() function.
- Try the following R code and see what happens. Intentionally the data x1 and x2 has been simulated here for demo. In reality, those x1 and x2 are usually extracted from data frame.

```
N=100
x1=rnorm(N)
x2=1+x1+rnorm(N)
plot(0) plot(0,1)
plot(x1) plot(x1,x2) ### scatter plot x1 on the horizontal
axis and x2 on vertical axis
plot(x2~x1) ### the same as above but using the formula (x2 as
a function of x1)
```

13.1.1 Titles, Legends and Annotations Titles

The **main** gives the main title and **sub** the subtitle. They can be passed as argument of the **plot()** function or using the **title()** function, and **xlab** gives the name of the x axis and **ylab** the name of the y axis. Here is one example for practice.

```
N=100
x1=rnorm(N)
x2=1+x1+rnorm(N)
plot(x1,x2,main="This is main title", sub="This is sub title", xlab=
    This x-axis label",ylab="This is y-axis label")
    ###### The same can be done separately as follows
plot(x1,x2,xlab="This is x-axis label",ylab="This is y- axis label")
title(main="This is main title")
title(sub="This is subtitle")
```

The size of the text can be modified using the parameters **cex.main**, **cex.lab**, **cex. sub** and **cex.axis.** Those parameters denote a scaling factor, i.e. the value of the parameter multiplies the size of the text. If you choose **cex.main=2**, the main title will be twice as big as usual.

13.1.1.1 Legend

The **legend()** function helps to make the legend to the plot. The position can be **bottomleft**, **bottomright**, **topleft**, **topright** or exact coordinates. Try this code to display the legend.

```
N=100
x1=rnorm(N)
x2=1+x1+rnorm(N)
plot(x1,type="l",col=1,lty=1) ### lty for line type, col for using
    color code lines(x2,col=2,lty=2) legend("bottomleft",
legend=c("x1","x2"),col=1:2,lty=1:2)
```

13.1.1.2 Text in the Margin and in the Graph

The function **mtext()** puts some texts in the margin. The margin can be at the bottom (side 1), the left (side 2), the top (side 3) or the right (side 4). For adding text to the graph itself, one can simply use the **text()** function. By default the space left in the margin in the number of lines. The default margins left are 5.1 in the bottom, 4.1 in left and top and 2.1 in the right. This can easily be changed by using **par (mar=c())**. For example, one can use **par(mar=c(4.1,4.1,4.1,4.1))** that leaves 4.1 lines in all the sides. The following example set the margin text and text in the graph itself.

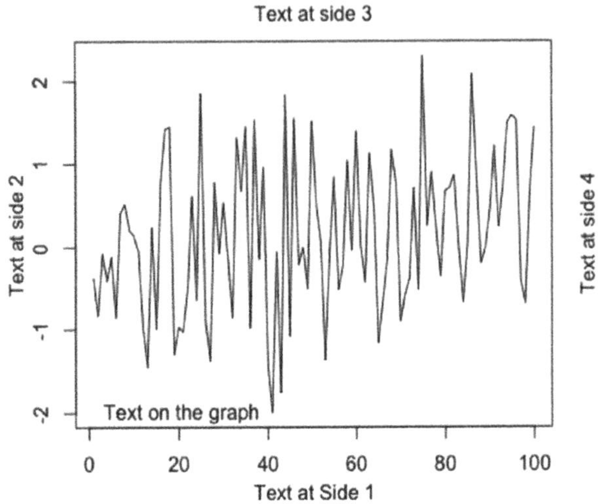

Fig. 13.1 Setting margin and margin text in R graphics

```
par(mar=c(4.1,4.1,4.1,4.1)) #### setting margins of all the
sides
x1=rnorm(100) plot(x1,type="l",col=1,lty=1,xlab= "",ylab="")
mtext("Text at Side 1",side=1,line=3)### at line 3 of the bottom
mtext("Text at side 2",side=2,line=2) ### at line 2 of side 2
(Leftside)
mtext("Text at side 3",side=3,line=2) ### at line 2 of side
3 (top)
mtext("Text at side 4",side=4,line=1) #### at line 1 of side
4 (Right)
   ####### Text inside the graph
text(20,-2, " Text on the graph") #### The text is at the
location (20,-2)
```

The graph shown in Fig. 13.1 is generated by running the above code.

13.1.1.3 Annotations and Customization

Further lines, points, arrow, segment and text can be added into the existing graph by using **lines(), points(), arrows(), segments()** and **text()** functions.

We can add mathematical symbols using the **expression()** function and make some substitution in a formula using **substitute()**. For knowing more about mathematical annotations with examples, one can seek help from the **help(plotmath)** command. Here is one example code. Try this code and see what happens.

```
xl=rnorm(100)
plot(xl,type="l",        ylim=c(-4,4),ylab="xl",xlab=expression
(paste
    ( "Temperature(","  "^0,"C)")))
text(80,-2,   expression(paste(frac(1,sigma*sqrt(2*pi)),    "",
plain(e)^{ frac(-
    (x-mu)^2,2*sigma^2)})))
```

13.1.1.4 Types of Plots

The type of a plot can be **n** for none (nothing is printed), **p** for points, **l** for lines, **b** for both, **o** for both overlaid, **h** for histogram-like and **s/S** for steps (Fig. 13.2).

```
par(mfrow=c(2,2)) ###### creating 4 graphs in a single page
x<-rnorm(100)
plot(xl,type="p", pch = 1, main="points", xlab= "",ylab= "")
### For different pch values seek help by issuing command help
("pch")
plot(xl,type="l", lty= 1,main="line",xlab= "",ylab="")
plot(xl,type="b",pch=1,lty=1,main="both",xlab= "",ylab= "")
```

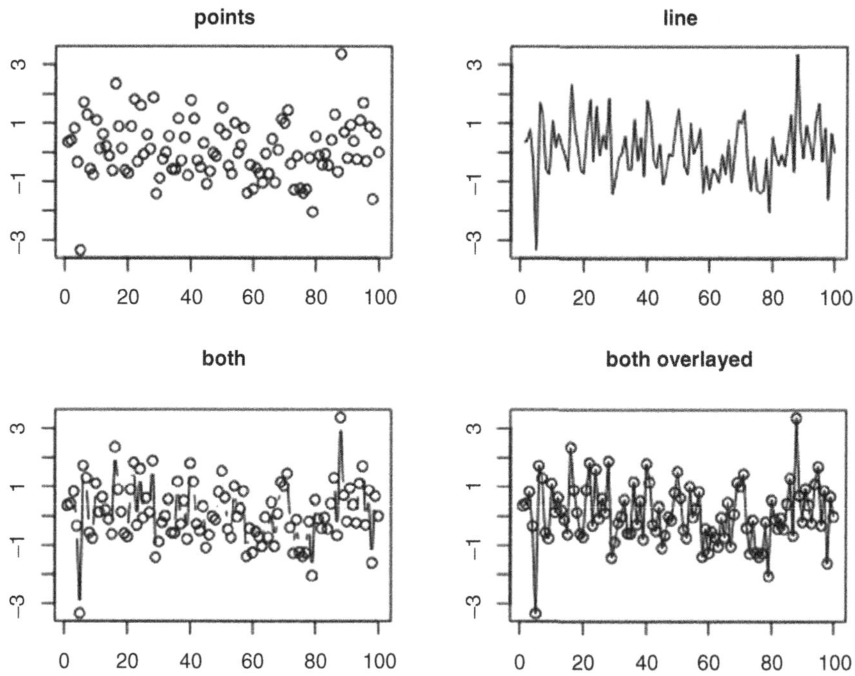

Fig. 13.2 Types of graph produced by plot function, for different values of type argument

13.1.1.5 Axes

The default output prints the axes. We can remove them by passing the argument **axes=FALSE** in **plot()** function. We can also change them using the **axis()** function.

Try out this code and see the results.

```
N=100
x1=rnorm(N)
x2=1+x1+rnorm(N)
plot(x1,x2,axes=FALSE)
axis(1,col="red",col.axis= "blue",font.axis=3)
axis(2,col="green",col.axis="magenta",font.axis=2)
```

13.1.1.6 Axis Label Style

The **las** value specifies the style of axis labels. It can be 0, 1, 2 or 3.

0: always parallel to the axis [default]
1: always horizontal
2: always perpendicular to the axis
3: always vertical

Try out the following R-code and get the graphic output as shown in Fig. 13.3.

```
x1=rnorm(100)
plot(x1,las=0,main="las = 0", sub = "always parallel to the axis",
    xlab="", ylab "")
plot(x1,las=1,main="las = 1", sub = "always horizontal", xlab="", ylab= "")
plot(x1,las=2,main="las = 2", sub = "always perpendicular to the axis",
    xlab= "",ylab= "")
plot(x1,las=3,main="las = 3", sub = "always vertical",xlab= "",ylab= "")
```

It is also possible to add another *y* axis on the right by adding **axis(4,)**. The following example shows the two *y*-axes to graph the catch and effort data on the same graph (Fig. 13.4).

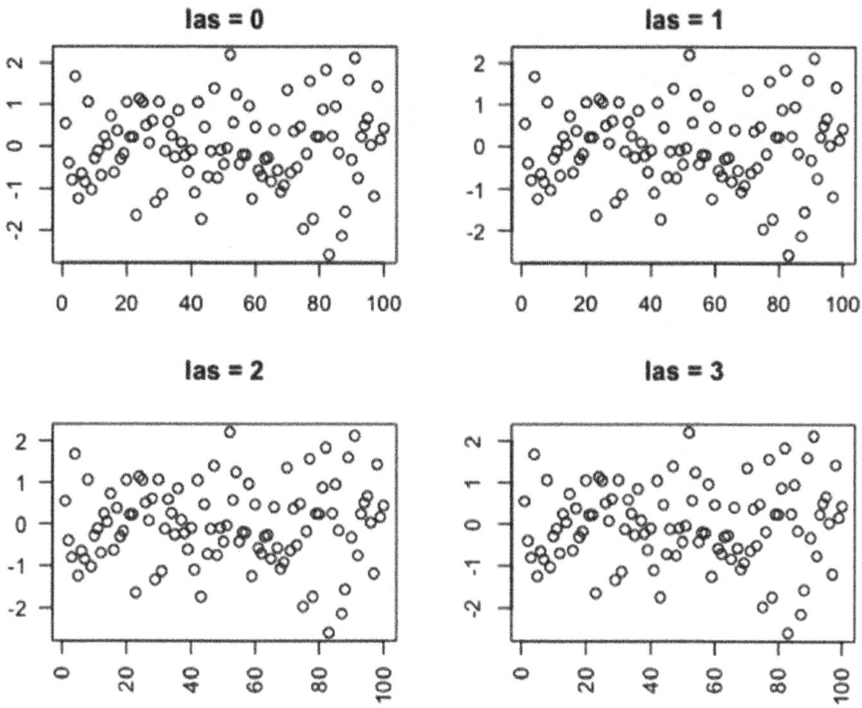

Fig. 13.3 Axis style change using 'las' argument

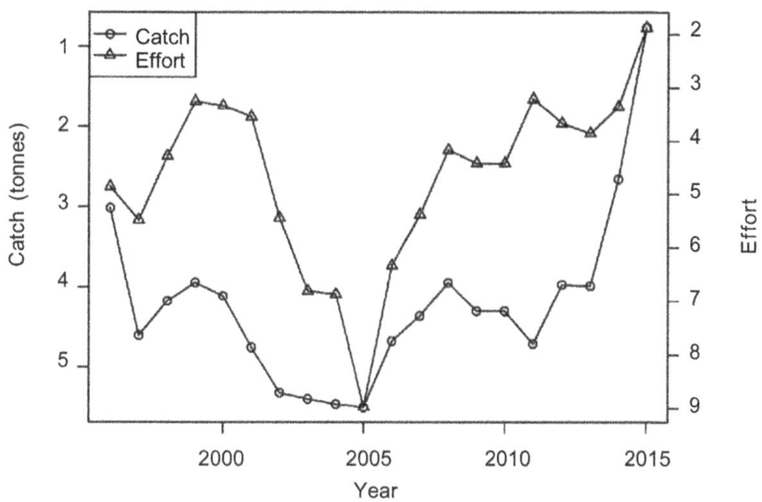

Fig. 13.4 The graph with two *y*-axes, representing catch and effort, respectively

```
dat=read.csv(file.choose(),header=T) ##### importing csv file
x1=dat$Year #### Extracting Year and assigning to x1
x2=dat$tcatch ### Extracting tcatch and assigning to x2
x3=dat$Effort ### Extracting tcatch and assigning to x3
par(mar=c(4.1,4.1,4.1,4.1)) #### setting margins of all the
sides
plot(x1,x2,type="o",       pch=1,lty=1,xlab="Year",ylab="Catch
(tonnes)") ### plotting
x1, x1 without axes title
par(new = TRUE) #### Setting for new graph
plot(x1,x3,axes=FALSE,type="o",pch=2,lty=2,xlab="",ylab="")
#### with no axes and
axes label
axis(4) #### setting 2nd y-axis
mtext("Effort",side=4,line=3) ### setting axis title by margin
text
legend("topleft", c("Catch", "Effort"),pch=1:2,lty=1:2) ####
setting legend
```

13.1.1.7 Colours

The colour of the points or lines can be changed using the **col** argument, **fg** for foreground colours (boxes and axes) and **bg** for background colours. The list of all the colours in R can be seen by using **colors()**. Try the following functions at R prompt and see the results. Note that x1 and x2 are the same as in the above example of catch and effort data.

```
colors() #### to see the list of colours
plot(x1,x2,type= "o",col= "red",
col.axis= "dodgerblue",
col.lab="firebrick", col.main="darkgreen",
col.sub="cyan4",main="Testing colors",
sub="Catch trend",ylab="Catch",xlab="Year")
```

We can also generate new colours using the **rgb()** function. The first argument is the intensity of red, second is the intensity of green, and third is the intensity of blue. They vary between 0 and 1 by default, but this can be modified with the option **max=255. col2rgb()** returns the RGB code of R colours. **col2grey()** and **col2gray()** convert colours to grey scale. Try following the code.

```
>newcolor=rgb(0.6,0.4,0.6)
>x1=rnorm(100)
>plot(x1,col=newcolor)
```

13.1.1.8 Plotting Curve

The **curve()** function plots functions. This can be added to an existing plot with the option **add=TRUE.** You can also plot functions using the **plot()**. The following example from real data demonstrates the **curve()** added with the existing plot (Fig. 13.5).

```
dat=read.csv(file.choose(), header=TRUE) ##### Importing CSV file
L=dat$Length ##### Extracting Length data
W=dat$Weight ##### Extracting Weight data
plot(L,W,xlab="Length(mm)", ylab="Weight(gm)") #####
curve(exp(-13.2075)*x^3.3016,min(L), max(L), add=TRUE,col="red",
lwd=2)
```

13.2 Statistical Graphics Gallery

In this section, we review on frequently used statistical plots drawn using R. This includes code for the standard graphics package, the **lattice** package and the **ggplot2** package. Make sure that **lattice** and **ggplot2** package is installed in your system and loaded in R before using the graphics functions. The above two packages are loaded by following two commands. This is the general method for loading any packages.

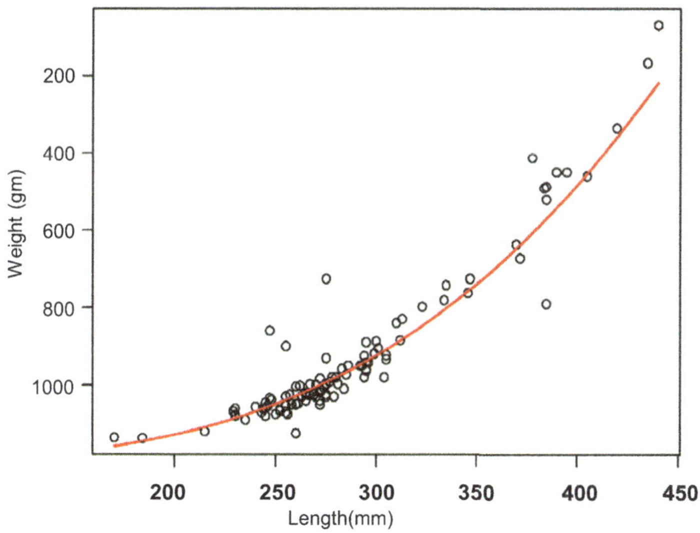

Fig. 13.5 A fitted curve added to the scatter plot

```
>library(lattice)
>library(ggplot2)
```

13.2.1 Line Plot

To draw a line plot, use the generic **plot()** function by setting **type="l"**. Then further lines on the same plot can be drawn using **lines()** function.

13.2.2 Scatter Plot

- One can invoke scatter plots using:
- Plot(x,y)—in standard graphics
- Plot(y x)—in standard graphics
- xyplot(y x)—in lattice
- qplot(x,y)—in ggplot2

13.2.2.1 Logarithmic Scale

Sometimes it is useful to plot the log of a variable and to have a log scale on the axis. It is possible to plot the log of a variable using the log option in the plot() function.

- For a log plot, use log = "xy".
- For a log in the *x* axis only, use log = "x".
- For a log in the *y* axis only, use log = "y".
- >plot(x,y, log = "xy") ### log-log plot

13.2.3 Scatter Plot Smoothing

Smoothing is a useful tool to examine the predictive pattern of relationship between two data series. The theoretical background is rather difficult for non-statisticians. R provides a simple function to play around the smoothing of scatter plot. The **scatter.smooth()** or **loess.smooth()** can be used to smooth the scatter plot. Other functions such as **loess()** and **lowess()** can also be explored. The readers may seek help on those functions.

Note *x1 and x2 have been simulated here. The users have to extract from data frame for their own data. The readers may practise it for length-weight data* (Fig. 13.6).

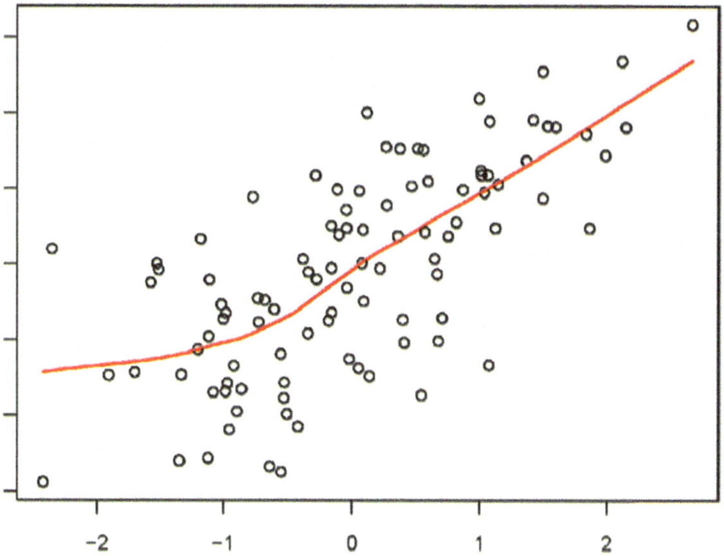

Fig. 13.6 Scatterplot smoothing to explore relationship

```
######### Scatter plot smoothing ####
x1=rnorm(100)
x2=1+x1+rnorm(100)
plot(x1,x2)
lines(loess.smooth(x1,x2),lty=2,lwd=2,col="red")
```

13.2.4 Box Plots

Box plot is usually used to see the variation of any numeric variable (e.g. weight, length, plankton abundance, fish abundance, etc.) over different factor levels. For example a researcher would be interested to plot variation of weight of fish collected over different places or month. Here 'box plot' is suitable.

Boxplots can be created for individual variables or for variables by group. The format is **boxplot(x, data=),** where *x* is a formula and **data=** denotes the data frame providing the data. An example of a formula is *y* group where a separate boxplot for numeric variable *y* is generated for each value of group. Add **varwidth=TRUE** to make boxplot widths proportional to the square root of the samples sizes. Add **horizontal=TRUE** to reverse the axis orientation.

```
boxplot(Weight~Place,data=dat,xlab="Place",ylab="Weight(gm)")
boxplot(Weight~Place, data=dat,xlab="Place",ylab="weight(gm)")
```

(continued)

```
boxplot(Weight~Place, data=dat,xlab="Place",ylab="weight(gm)",
varwidth=TRUE)
#### width related to sample size
boxplot(Weight~Place, data=dat,xlab="Place",ylab="weight(gm)",
horizontal=TRUE)
```

13.2.5 Histogram Plot

A histogram is a plot of the counts or proportion of data in disjoint intervals along the data set. The actual values of the data are not used; a datum's presence in a particular interval (or 'bin') is tallied, and a histogram displays vertical bars with height representing the tally or the proportion of the sample size in each bin. Histogram plot is extremely useful for visualizing the distribution pattern of a variable. Specifically, it is useful to examine:

- Whether the distribution is symmetric, left skewed or right skewed
- How many modes does the distribution have and if there are there any outliers

Histogram plot is invoked by **hist()** function. Use argument **prob=TRUE** to display *y* axis in proportion or density. In the following example, we will examine the fish weight distribution.

```
x=dat$Weight #### Extracting Weight from dataframe 'dat' and
assigning to x
hist(x,xlab="Weight(gm)",   main="Weight   Distribution   of
Hilsa") #####y=axis counts
hist(x,xlab="Weight(gm)", prob=TRUE,main="Weight Distribution
of Hilsa")
    #### y-axis density
y-axis density########### Histogram overlaid with Normal curve
hist(x,xlab="Weight(gm)", prob=TRUE,main="Weight Distribution
of Hilsa")
    #### y-axis density
m=mean(x) ##### computing mean of Weight
std=sd(x) #### Compute sd of Weight
curve(dnorm(x,mean=m,sd=std),min(x),max(x),add=TRUE,
col="red") #### overlaying
    normal distribution
box()
```

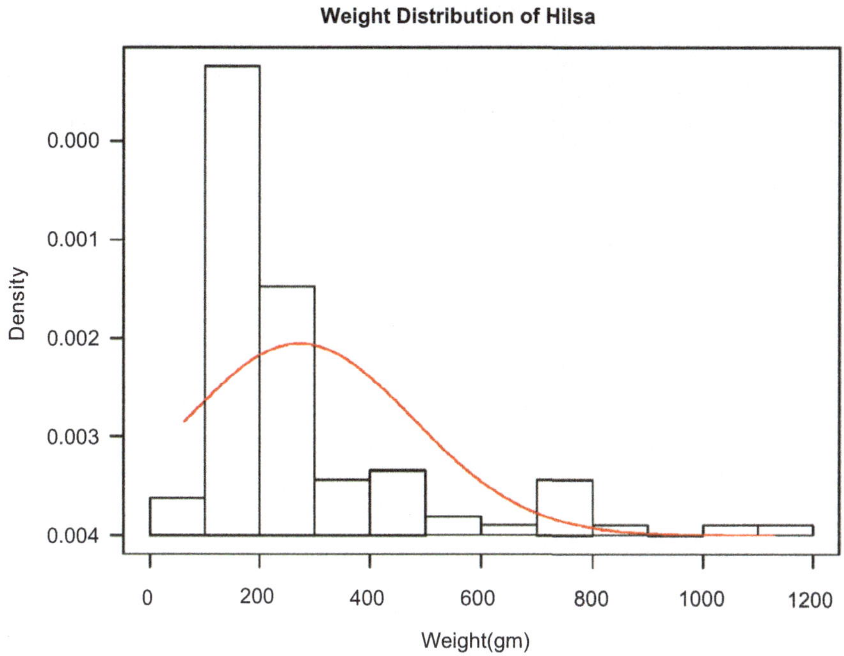

Fig. 13.7 Histogram overlaid by normal density

Interpretation

Weight distribution (see Fig. 13.7) of fish is highly skewed and deviates from normality, and there is possibly one outlier. Plausible explanation is that the fishes were caught using particular mesh size.

13.2.6 Histogram with Density Estimate

```
hist(x,xlab= "Weight(gm)", prob=TRUE,main= "Weight Distribution of
    Hilsa") #### y-axis density
d=density(x) #### density estimate
lines(d,col= "red") #### overlaying density estimate
```

Interpretation

The weight distribution (see Fig. 13.8) has three modes. There is a possibility of using three different mesh sizes inducing three modes in the weight distribution.

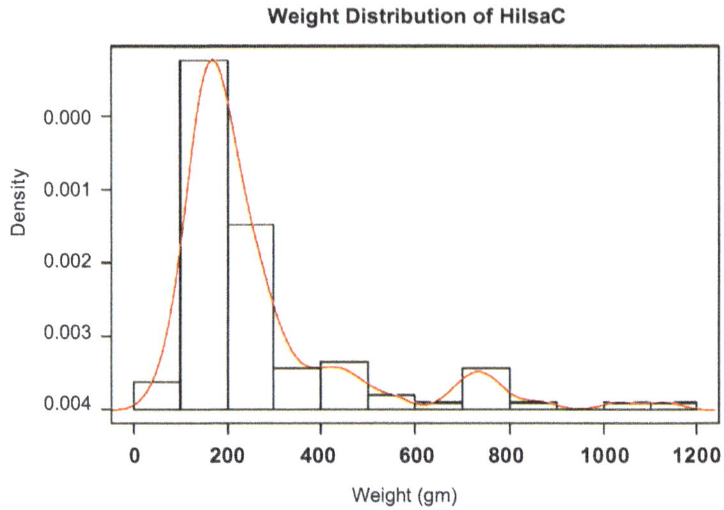

Fig. 13.8 Histogram with overlaid kernel density estimate

13.2.7 Normal Q-Q Plot

Q-Q plot is useful to examine the normal distribution of a data series. If the data series is normally distributed, observation should be approximately on a straight line. Otherwise, if the data is non-normal, the points deviate markedly from a straight line. Possible outliers are points at the ends of the line, distanced from the bulk of the observation. The **qqnorm()** and **qqlines()** can be used to generate Q-Q plot.

```
qqnorm (dat$Weight)
qqline (dat$Weight,col = "red")
```

Note

Make sure the data frame has already been imported and stored in the variable 'dat' wherein 'Weight' is a column.

Interpretation It is evident from Fig. 13.9 that the distribution deviates from normal distribution, as the large number of points lies outside the straight line.

Ideally, all the points should lie on the line for perfect normally distributed data.

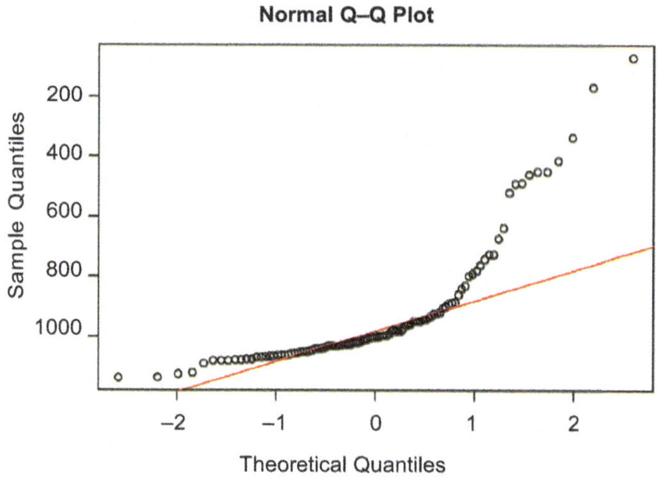

Fig. 13.9 Q-Q plot to examine the normality. This example is for the weight distribution of Hilsa

13.2.8 Empirical Distribution Function Plot

The empirical cumulative distribution function (commonly known as CDF) is a nonparametric estimator of the underlying CDF of a data series. Essentially it assigns a probability of $1/n$ to each datum, orders of the data from smallest to largest in value. It calculates the sum of the assigned up to and including each datum. The empirical CD is usually denoted by

$$F(x) = Pn(X \leq 5x) \tag{13.1}$$

It can be conceptualized as the probability (or proportion in practical sense) of a datum less than equal to x. It can be interpreted in terms of proportion as well. The following example illustrates the plot which is generated using **ecdf()** function.

```
>cdf=ecdf(dat$Weight)
>plot(cdf, main="CDF of Weight of Hilsa,xlab= "Weight(gm)")
>print(ecdf(500)) ####### cdf at 500 gm
[1] 0.79028
>abline(v=500) ##### adding a vertical line at 500 gm
```

Note

Make sure the data frame has already been imported and stored in the variable 'dat' wherein 'Weight' is a column.

The graph shown in Fig. 13.10 is obtained by using the above R-code.

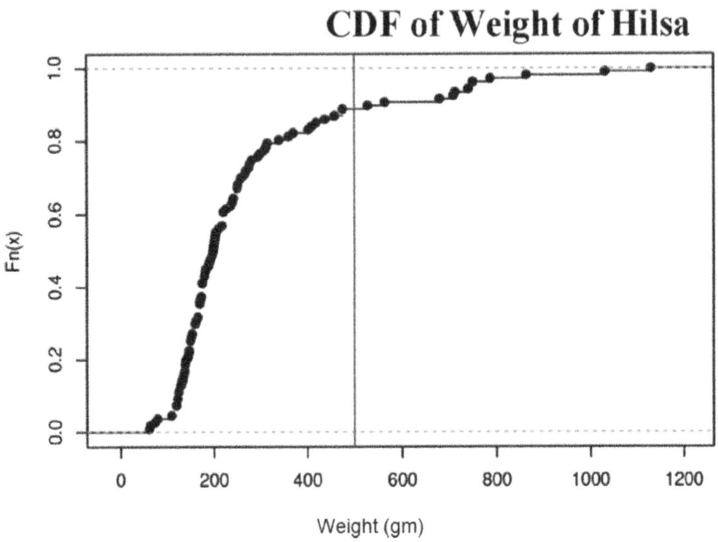

Fig. 13.10 Empirical cumulative distribution function for the weight distribution of Hilsa

Interpretation Proportion of larger fish caught is small (see Fig. 13.10). Visual inspection indicates that proportion of fish below 500 gm is 0.79028 (approx. 79%).

This proportion can be computed or visually inspected at any size.

13.2.9 Bivariate Contour Plot

It is useful when there are three variables, and one variable varies over two other variables. The contour plot may be used to explore the data.

The following example R-code uses the contour plot (Fig. 13.11).

```
>dat<-read.csv("breeding-ground.csv",header=T) ### Importing
.csvfile
>dat<-na.omit(dat) ####### omitting blank data
>depth<-dat$depth*0.305 +0.001 #### extracting depth
>distance<-dat$cumdist+0.001 ###### extracting distance
>GSI<-dat$GSI+0.001 #### Extracting GSI
>fit.GSI<-loess(GSI~depth*distance,parametric=c("distance"),
span=2,
    degree=2) #### Loess smooth
>depth.marginal<-seq(min(depth),max(depth),length.out=100)
```

(continued)

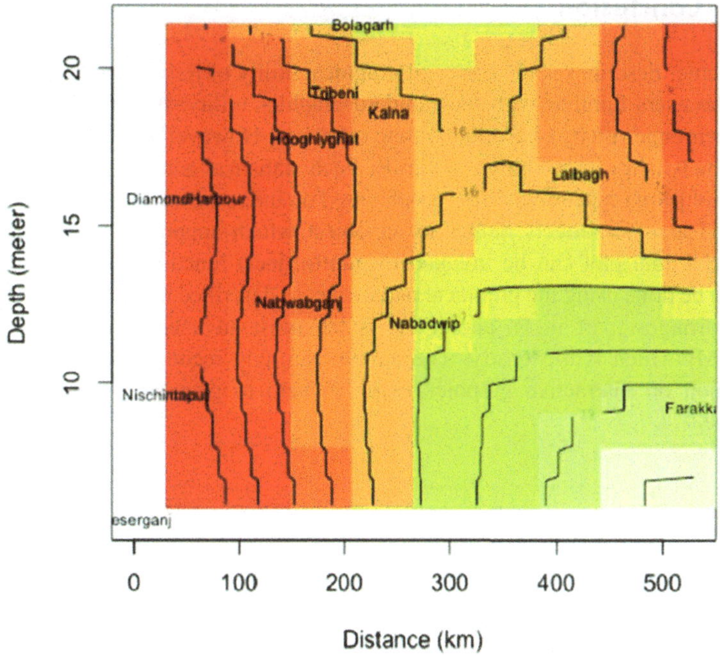

Fig. 13.11 Contour plot for identification of breeding ground of Hilsa

```
#### setting
    y-axis marginal
>dist.marginal<-seq(min(distance),max(distance),length.
out=10) ### Setting x-axis
>dd.marginal<-list(distance=as.integer(dist.marginal),
depth=as. integer(depth.marginal))
### preparation for
>grid<-expand.grid(dd.marginal) #### Creating grid
>fit<-predict(fit.GSI,grid) ##### predicting over grid
>plot(distance,depth,type="n",xlab="Distance(km)",ylab="Depth
(meter)") #### plotting
the outline
>image(dist.marginal,depth.marginal,fit,add=T)   ####   placing
image
>contour(dist.marginal,depth.marginal,fit,add=T)   ###   Adding
contour
>text(distance,depth,dat$CenterName,cex=0.7) ### adding text
```

13.3 Conclusion

This chapter describes some basics of graphical utility of R. The intention here is to make the users familiar with the standard graphics utility. Interested readers may build their expertise by consulting online resources. However, there are huge variety of graphs which can be drawn within R. Each statistical analysis provides generic support of **plot()** function to draw graph based on the results obtained. Usually, each statistical analysis directs output to an object which supports the generic **plot()** function. Initial plot can be invoked by **plot(object)** function. Further tweak and tune can be done using the graphical parameters as described in the section. Graphics is the strongest part in R, and it offers the most advanced annotation next to MATLAB which is too costly. The authors strongly recommend to build strong foundation on interactive graphics that will help to generate graph which they visualize.

Exercise

1. Which of the following packages is not related graphics in R?
   (a) ggplot2
   (b) Lattice
   (c) MASS
   (d) Trellis
2. R graphics is
   (a) Interactive
   (b) Non-interactive
3. Which is the correct command to view demo graphics in R?
   (a) View (demo)
   (b) Demo(graphics)
   (c) Demo(plot)
   (d) None of the above
4. Which of the following is not a graphics function in R?
   (a) text()
   (b) points()
   (c) lines()
   (d) mean()
5. What is the generic function to produce a graph in R?
   (a) graph()
   (b) plot()
   (c) figure()
   (d) None of the above
6. Suppose you have the catch series data of a reservoir for the Year = c(2001, 2002, 2003, 2004, 2005) and Catch = c(500, 600, 400, 700, 800). Which of the following is the command to generate a line graph?

(a) plot(Year, Catch)
(b) plot(Catch, Year)
(c) plot(Year, Catch, type="l")
(d) None of the above

7. Suppose you want the *y* axis label as 'Catch (MT)' in the exiting figure generated in Question 6. Which of the following command is true?
(a) plot(Catch, Year, type="l",ylab="Catch(MT)")
(b) plot(Year, Catch, type="l", label="Catch(MT)")
(c) plot(Year, Catch, type="l", ylab="Catch(MT)")
(d) plot(Year, Catch, type="l", xlab="Catch(MT)")

8. For the same data as in Question 6, what will be the command if you wish to generate point only graph instead of line graph?
(a) plot(Year, Catch, type="p")
(b) pointgraph(Year, Catch)
(c) plot(Year, Catch, point)
(d) None of the above

9. For the same data as in Question 6, what will be the command if you wish to generate with both line and point?
(a) plot(Year, Catch, type="lp")
(b) plot(Year, Catch, type="b")
(c) linepoint(Year, Catch)
(d) plotlinepoint(Year, Catch)

10. Which command is used to annotate different title of an existing graph?
(a) title()
(b) Title ()
(c) changeTitle()
(d) None of the above

11. Which of the following is incorrect to get help on the command **plot()** function?
(a) ?plot
(b) help("plot")
(c) Both of the above
(d) HELP(plot)

12. Which is the correct command to place a legend in an existing graph?
(a) graphlegend()
(b) Legend ()
(c) L()
(d) legend()

13. Which of the following is the correct command for drawing a bar chart?
(a) bargraph()
(b) barchart()
(c) barplot()
(d) None of the above

14. Which of the following is the correct command for drawing a line chart?
(a) linechart()
(b) plot()

   (c) linegraph()
   (d) None of the above
15. Which of the following function is used to draw box-whisker graph?
   (a) boxplot()
   (b) box-whisker()
   (c) BOX()
   (d) None of the above
16. Using the data given in Table 3.3, reproduce the Fig. 3.1 in R.
17. Generate a bar chart in R using the data given in Table 3.4.
18. Draw a histogram in R using the data given in Example 3.1.
19. Which of the following command is correct to draw a horizontal line at 0 of y axis in an existing plot?
   (a) lineat(h=0)
   (b) horizline(h=0)
   (c) abline(h=0)
   (d) None of the above
20. Q-Q plot is extremely useful to examine the normal distribution of a data series. What is the correct command to draw a Q-Q graph?
   (a) Qplot()
   (b) Qnorm()
   (c) Q-Qplot()
   (d) qqnorm()

Further Reading

Maindonald J, Braun J (2003) Data analysis and graphics using R: an example-based approach. Cambridge University Press, Cambridge ISBN 0-521-81336-0. http://wwwmaths.anu.edu.au/~johnm/

Murrell P (2005) R graphics. Chapman & Hall/CRC, Boca Raton, FL ISBN: 1-584-88486-X. http://www.stat.auckland.ac.nz/~paul/RGraphics/rgraphics.html

Appendix: Answers of All Exercises

Chapter 2

1. $\hat{X} = 125$; Margin Error = 65.5; Confidence Interval = (59.5 to 190.5)
2. $\hat{X} = 1860$ kgs; SE($\hat{X}$) = 310.913
3. $\hat{X} = 6600$ kgs; SE($\hat{X}$) = 273.7371 kgs
4. (i) x = Rs. 3378; SE($\hat{X}$) = 254.731; (ii) ($\hat{X}$) = Rs. 101364; SE($\hat{X}$) = 7641.929

Answer of MCQ

5. c; 6. d; 7. b; 8. c; 9. d; 10. d

Chapter 3

1. a; 2. b; 3. d; 4. d; 5. c; 6. a; 7. b; 8. a; 9. d; 10. a;
11.– 12.– 13. d; 14. a; 15. b; 16. c; 17. c; 18. a; 19. c; 20. b

Chapter 4

1. Accept H_0: the sample is drawn from a population with mean 8.1 and standard deviation 0.25 cm.
2. 9.95 _ 0.433 or (9.52, 1038) cm.
3. Accept H_0: Proportion do not differ significantly.
4. Reject H_0: The inference is that the fish feed purchased at online auctions is less than Rs. 179.
5. Reject H_0: Training led to improvements.
6. Reject H_0: The means differ.
7. c
8. a

B. K. Das et al., *Concept Building in Fisheries Data Analysis*,
https://doi.org/10.1007/978-981-19-4411-6

9. $z = (X–\lambda)/\rho \sim N(0,1)$
10. Power $= 1–\alpha = P(\text{Reject } H_0 | H_0 \text{is false})$

Chapter 5

1. a; 2. b; 3. b; 4. a; 5. b; 6. a; 7. a; 8. d; 9. c; 10. d;
11. c; 12. b; 13. a; 14. b; 15. c

Chapter 6

1. a; 2. a; 3. b; 4. c; 5. a; 6. b; 7. d; 8. a; 9. a; 10. c;
11. d; 12. b; 13. a; 14. d; 15. b; 16. c; 17. b; 18. a; 19. a; 20. c

Chapter 7

1. c; 2. b; 3. a; 4. b; 5. c; 6. a; 7. d; 8. a; 9. b; 10 a;
11. d; 12. c; 13. c; 14. c; 15. b; 16. c; 17. d; 18. b; 19. c; 20. c

Chapter 8

1. c; 2. c 3. b; 4. a; 5. c; 6. b; 7. b; 8. c; 9. c; 10. a;
11. b; 12. c; 13. d; 14. b; 15. c; 16. c; 17. a; 18. d; 19. a; 20. a

Chapter 9

1. c; 2. b; 3. d; 4. c; 5. b; 6. a; 7. a; 8. d; 9. b; 10. d

Chapter 10

1. c; 2. a; 3. b; 4. d; 5. a; 6. c; 7. c; 8. d; 9. b; 10. a;
11. a; 12. b; 13. d; 14. a; 15. c; 16. a; 17. b; 18. d; 19. c; 20. d;
21. a; 22. b; 23. b; 24. c; 25. b

Chapter 11

1. d; 2. b; 3. a; 4. c; 5. c; 6. d; 7. a; 8. a; 9. d; 10. a;
11. b; 12. c; 13. d; 14 b; 15. c; 16. d; 17. b; 18. c; 19. d; 20. b

Chapter 12

1. a; 2. b; 3. c; 4. c; 5. b; 6. b; 7. a; 8. c; 9. a; 10. b;
11. d; 12. c; 13. a; 14. d; 15. d; 16. c; 17. a; 18. d; 19. a; 20. d

Chapter 13

1. a; 2. a; 3. b; 4. d; 5. b; 6. c; 7. c; 8. a; 9. b; 10. a;
11. a; 12. d; 13. c; 14. b; 15. a; 16. –; 17. –; 18. –; 19. c; 20. d

The manufacturer's authorised representative in the EU is Springer Nature Customer Service Centre GmbH, Europaplatz 3, 69115 Heidelberg, Germany. If you have any concerns regarding our products, please contact ProductSafety@springernature.com

Printed and bound by CPI Group (UK) Ltd, Croydon, CR0 4YY
30/09/2026
02225242-0006